GW01402678

A2 Geography
UNIT 5

Edexcel

Specification **A**

Unit 5: Human Systems, Processes and Patterns

Nigel Yates

Philip Allan Updates
Market Place
Deddington
Oxfordshire
OX15 0SE

tel: 01869 338652
fax: 01869 337590
e-mail: sales@philipallan.co.uk
www.philipallan.co.uk

This Guide has been written specifically to support students preparing for the Edexcel Specification A A2 Geography Unit 5 examination. The content has been neither approved nor endorsed by Edexcel and remains the sole responsibility of the author.

Printed by Raithby, Lawrence & Co. Ltd, Leicester

Contents

Introduction

■ ■ ■

Content Guidance

■ ■ ■

Questions and Answers

Introduction

About this guide

This guide is for students following the Edexcel Specification A A2 Geography course. It aims to guide you through Unit Test 5, which examines the content of **Unit 5: Human Systems, Processes and Patterns**.

This guide will clarify:
- the content of the unit so that you know and understand what you have to learn
- the nature of the unit test
- the geographical skills and techniques that you will need to know for the assessment
- the standards you will need to reach to achieve a particular grade
- the examination techniques you will require to improve your performance and maximise your achievement

This **Introduction** reviews the main features of the unit and the approaches to studying it. It also explains some of the techniques commonly used in writing good essays, with coverage of command words and some advice on how to construct examination answers.

The **Content Guidance** section summarises the essential material for Unit 5. The aim is to make you aware of the type of information and understanding of abstract concepts that you will need to make sense of the facts.

The **Question and Answer** section presents essays drawn from each section of the unit. Each answer includes a detailed analysis from a senior examiner, illustrating both the positive and the negative aspects of the essay.

A2 Geography

The A2 course is designed for those who have already completed AS Geography. It has different units from the AS course and is intended to be more demanding. The course will enable you to combine your AS results with your A2 results to receive an A-level grade.

Scheme of assessment

Unit 5: Human Systems, Processes and Patterns is one of three units that make up the A2 specification. The A2 element is worth 50% of the full A-level qualification. In this specification, there are no options as to which units you can study; all candidates must take Units 4, 5 and 6.

The Unit 5 test is marked out of 50, but you are awarded a uniform mark for the paper, which is out of 90. Although you are not awarded a grade for each unit, the marks awarded are equivalent to the following grades: 72–90 = A; 63–71 = B; 54–62 = C; 45–53 = D; 36–44 = E; < 36 = U.

Unit	Unit test length	Max. mark	Max. uniform mark	A2 weighting
4: Physical Systems, Processes and Patterns	1 hour 30 minutes	50	90	30%
5: Human Systems, Processes and Patterns	1 hour 30 minutes	50	90	30%
6: Synoptic	2 hours	75	120	40%

Unit 5

The specification content of Unit 5 comprises three sections:
- Economic systems
- Rural–urban interrelationships
- Development processes

These are studied at a range of scales from local to global. You need to appreciate the interaction of human activity with the physical environment and there is a requirement to study specific places and examples.

During the course you should aim to:
- develop a knowledge of geographical terminology, concepts, principles and theories
- acquire and apply knowledge and understanding of physical processes, their interactions and outcomes over space and time, through the study of places and environments
- acquire and apply a range of geographical and transferable skills necessary for the study of geography, particularly to describe, analyse and interpret data and resources
- develop an understanding of the relationships between physical environments and people
- appreciate the dynamic nature of geography — how places and environments change

Revision advice

It is important that you keep up to date with your studying and complete all work set on time. If you are absent for any reason, you need to find out what you missed and catch up as quickly as possible. Hopefully, you will have some time for revision at the end of the course and you should try to make full use of any opportunities that you are given.

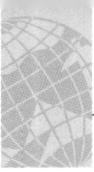

When you are revising on your own, try to do the following:
- Plan a revision schedule that gives fairly even time allocation to each of your subjects and to each unit within the subject.
- Stick to the schedule!
- Read through all material relating to a topic — notes, handouts, questions, worksheets, etc.
- Make revision notes that summarise the key elements of the topic (the Content Guidance section of this book will help).
- Make a list of the key facts/figures for each of the main case studies/examples that you need.
- Practise sample/past examination questions (such as the ones in the Question and Answer section).
- Ask your teacher/lecturer to clarify any topics/issues about which you are uncertain.
- Remember, you need to revise *all* of the specification content in this unit.

Examination skills

For each topic, you will be assessed on your knowledge, understanding and skills. The examination requires you to write *two* answers from the six questions on the paper in 90 minutes; two questions are set on each of the three unit sections. You must answer from different sections.

You may have studied all three sections or just two sections. If you have taken the first route, you will have a wider choice of questions on the examination paper. If you have taken the second route, although limited to choosing two out of four, you should have more depth to your knowledge and understanding.

Each question carries 25 marks. The first part of the question is for 5 marks, the second part for 20 marks.

Each question has a resource. This could be:
- a graph
- a map
- a table
- a schematic diagram
- a photograph
- a cartoon

The resource is there to guide you to the correct area of unit content for your answer. There may be some useful information within the resource — a few hints about the relevant topic, perhaps — but *you are not required to manipulate the resource* in any way, or to describe it or to mark anything on it. It is there only as a stimulus.

The 5-mark part of the question usually asks you for an extended definition of an idea or theme, or asks you to compare two ideas. The 20-mark part of the question is the essay question. This is drawn from a recognisable part of the specification content.

The wording of some questions will be familiar, whereas other questions may approach a topic from a less conventional angle.

Choice of questions

The questions in the paper are in the same order as the specification: questions 1 and 2 are on economic systems; questions 3 and 4 on rural–urban interrelationships; and questions 5 and 6 on development processes.

The first task is to choose the right questions to answer. Some students make a false start to an answer and then abandon it, losing valuable time. You should know what to expect and be familiar with the shape and structure of the paper.

- Read all the questions (even those on sections of the specification that you do not think you have been taught).
- Do not spend too much time on the first, 5-mark part of the question. It matters, of course, but not as much as the 20-mark part. Look carefully at the question and, using the techniques suggested here, establish whether you might be able to answer it.

If you have no choice on one section of the question paper, but are undecided about your second essay, then get on with your first. While you are doing this, you may clarify your ideas about the second essay, even subconsciously.

Watch out for titles that are similar, but not identical, to titles that you might have written during the course. Write an answer to the question set and not a prepared answer to a title that seems roughly the same.

Understanding the question

Students often achieve a low mark in an examination essay because they have not fully understood the question. To score high marks in an examination or an essay, it is important to understand how the question should be answered. In order to do this, it is useful to analyse the question and identify its components and command words.

Components of a question

Most essay titles or examination questions contain the following components:

- subject matter or topic — what, in general terms, is the question about? From what section of the specification is it drawn?
- aspect or focus — this is the angle or point of view on the subject matter. What aspect of the subject matter is the question about?
- command or instruction — this refers to the command word or phrase. This tells the student exactly what to do.
- restriction or expansion of the subject matter — this is the detailed limitation of the topic. What, in specific terms, is the question about? Is it all industry or just manufacturing industry?
- viewpoint — you may be required to write from a viewpoint dictated by the question. Such a question could begin with 'Criticise the view that…'

Command words

The questions use a series of command words that you need to understand. The list below explains the meaning of the commonly used command words.

- **Account for:** give the reasons for the subject of the question.
- **Analyse:** take apart an idea, concept or statement, in order to consider all the factors of which it consists. Answers of this type should be very methodical and logically organised.
- **Compare:** set items side by side and show their similarities and differences. A balanced, objective answer is expected.
- **Consider:** describe and give your thoughts on the subject.
- **Contrast:** point out only the differences between two items.
- **Criticise:** point out mistakes or weaknesses, but *also* indicate any favourable aspects of the subject of the question.
- **Define:** explain the precise meaning of a concept. This should include recognition of any difficulties in defining the term.
- **Describe:** say what something is like, how it works and so on.
- **Discuss:** explain an item or concept, and then give details about it, using supportive information, examples, points for and against, and explanations for the facts put forward. It is important to give both sides of an argument and come to a conclusion.
- **Elucidate:** explain what something means and make it clear (lucid).
- **Evaluate/assess:** decide and explain how great, valuable or important something is. The judgement should be backed by a discussion of the evidence or reasoning involved.
- **Examine:** investigate in detail, offering evidence both for and against a point of view or a judgement. Both description and explanation are likely to be required, and some sort of personal observations or comments are expected.
- **Explain:** offer a detailed and exact explanation of an idea or principle, or a set of reasons for a situation or attitude.
- **Explore:** examine the subject thoroughly and consider it from a variety of viewpoints.
- **Illustrate:** provide examples to demonstrate or prove the subject of the question. This command is often added to another instruction.
- **Justify:** give only the reasons for a position or argument, and the main objections likely to be made to them.
- **State:** express the relevant points briefly and clearly, without lengthy discussion or minor details.
- **Summarise/outline:** provide a summary of all the available information about a subject. Questions of this type often require short answers.
- **Trace:** state and describe briefly, in logical or chronological order, the stages in the development of a theory, a settlement pattern, a process, etc.
- **To what extent is it true that...:** discuss and explain in what ways the statement is true and in what ways it is not true.

Analysing the question
- Identify the **topic** (put a circle around it).
- Recognise the **command** (underline it).
- Search for the **aspect** — this is the angle or point of view on the subject matter.
- If the topic has a **restriction** or **expansion**, identify it.

Answering the question

Writing an essay plan
You will have written essay plans while preparing for the examination and as a part of your course.
- Begin by highlighting or circling the key words in the title.
- Outline the key themes that you would like to address, under separate headings.
- Look very carefully at the resource, noting anything that might be useful in the construction of your answer.
- Refer to it in your answer should it seem appropriate to do so.

You should not spend more than 4 or 5 minutes doing this.

The introduction to the essay
The introduction should be a context statement, which says something meaningful. Clarify your terms through extending definitions or point out how some of the terms used in the question are difficult or problematic. Try to set the tone of the essay by stating how this topic interrelates with others.

The middle (analytical part) of the essay
The core of the essay should be organised in paragraphs, each of which takes on an aspect of the question, with examples. Structure it logically by tackling one point in each paragraph. Do not drift off into themes or evidence that are not what the question requires. Use evidence constructively and make it work for you. Remember that you can never lose marks. A bold guess is better than nothing. You may have only a hazy recall of the precise rate at which the South Korean economy grew in the 1980s, but better a hazy idea than none at all. If in doubt, include it!

Throughout this section, try to make linkages between the evidence offered and the title. There are a whole series of very useful words and phrases to help you do this:
- **concept:** an important idea
- **concise:** short, brief
- **in the context of:** referring to, inside the subject of
- **criteria:** what standards you would expect; what questions you would expect to be answered
- **deduction:** the conclusion or generalisation you come to after looking carefully at all the facts
- **factor(s):** the circumstance(s) bringing about a result
- **function:** what something does; its purpose or activities
- **implications:** results that are not obvious; long-term, suggested results

- **limitations:** where something is not useful or not relevant
- **with/by reference to:** about the following subject
- **in relation to:** only a certain part of the first topic is needed
- **role:** what part something plays, how it works, especially in cooperation with others
- **scope:** the area where something acts or has influence
- **significance:** the meaning and importance

In your analysis of the question, bear in mind the following:
- **contrasts in scale** — does your answer tackle the fact that some explanations work at some scales but not at others? Hence, climate might be the key variable in controlling agricultural land use at a global scale but not at a local scale.
- **contrasts in space** — does your answer allow for the fact that what is true for one part of the world may not apply to another? An example of this is the contrast between MEDCs and LEDCs.
- **contrasts in time** — does your answer allow for the fact that explanations will vary from one period to another? What was true in 1945 may not be true today. Hence, a theory such as Weber's may well be applicable in the circumstances of one period, but not another.

The end (conclusion) of the essay

This should *not* be a précis of the middle part. Start your conclusions by referring back to the original question and its key aspects or proposition. Try to leave the examiner with the impression that you could have said more if only you had been allowed the time. All conclusions in human geography are liable to be partial, tentative and incomplete, and you are perfectly entitled to say so.

How to structure an essay

An example of how to structure an essay on the question 'Describe and explain the causes and consequences of the changing location of manufacturing industry' is provided below.

Introduction
- **Define terms:**
  - manufacturing industry (distinguish between consumer and capital)
  - location (make a point about scales)

Make a general point about the significance of manufacturing.
Mention the increasing complexity of the global economy.

Main themes
- Case studies (especially automobiles and textiles)
- Early history of manufacturing — location determined by materials and linking industries
- Constraints of transport systems
- Labour skills
- Set in the context of classical location theory

- Development of industry — emergence of Fordism
- Increasingly global aspect to production
- Location changes at both scales — i.e. site factors and within a country
- Special nature of the UK's industry
- Post-war changes, especially in the UK
- The rise of Japan — address JIT and impact on location (local scale)
- Address entry of Japan into European market and US market and the impact on location (national and international scale)
- Rise of NICs — Korea/Malaysia
- Increasing globalisation of production
- Increasing rationalisation of industry
- Bringing cheap labour to the producer, e.g. from textiles

Conclusion
- Outline changes in location at global level
- Itemise main causes
- Itemise main consequences
- Make point about international division of labour

Content
Guidance

There are three sections in the specification content for Unit 5:

(1) Economic systems

(2) Rural–urban interrelationships

(3) Development processes

In this Content Guidance, each of the three elements is considered in terms of:
- the key concepts involved
- the content required
- the use of examples and case studies

This should make it clear to you what you need to know and understand.

Economic systems

Economic activity is highly varied in employment structure, organisation and location

The classification and characteristics of economic activity

Economic activity can be divided into primary, secondary, tertiary and quaternary sectors. This is a well-known sub-division of employment, but there are a number of common errors and simplifications, which need to be avoided.

Primary sector

The **primary sector** is fundamental because it produces the raw materials on which all other economic activity depends. Primary activity involves the exploitation of raw materials (coal mining or drilling for oil); the growth of food crops and textile crops in agriculture; fishing; and any other activity in which very little processing of the basic material is involved, for example quarrying. In these extractive processes, little value is added to the value of the product itself because little is done to it. Therefore, these are called **low value-added** industries.

Secondary sector

The **secondary sector** adds more value to the raw materials by processing them in some way or, more usually, both processing them and combining them to create goods. Thus, growing cotton is a primary industry, but the making of cloth and clothing is secondary. To make something requires labour and, almost always, some equipment or capital.

It is useful to distinguish between secondary industries that produce essential equipment for other industries (e.g. machine tools, iron and steel), which are often known as **capital** or **basic industries**, and those that produce goods for direct sale to consumers.

Consumer industries took on a leading role in economic growth in the twentieth century. In the last 30 years, the development of a distinctive group of **high-tech industries** has emerged, characterised by high levels of technological input and a heavy reliance on the development of new products that are faster, cleaner and more efficient.

Tertiary sector

The **tertiary sector** is sometimes referred to as the service sector. It produces no physical product; in other words, nothing that you can touch. Historically, the largest single group in the tertiary sector has been servants or slaves. The main categories today include fast food operatives and geriatric health care workers. In neither case do these people enjoy the high levels of wages that many associate, wrongly, with this sector. There are some very highly paid tertiary workers, but they are in the minority in all societies. Tertiary employment cannot exist alone because it always depends upon primary and secondary activity for the provision of the basic necessities of life.

Quaternary sector

The **quaternary sector** is a sub-set of the tertiary sector. It is a new category which recognises the importance of information technology and research in the creation of wealth. It includes employment that is involved with the gathering and exchange of information, as well as activities that concentrate on the development of ideas and research and senior management roles. The common confusion here is to include so-called 'high-tech' industry (which is the manufacturing of goods) with a highly advanced, usually electronic system of production. The production of microprocessors is a **high-tech** industry belonging to the secondary sector. However, there will certainly be people involved in the development of new products in this industry that take on quaternary activities.

These employment categories do not necessarily reflect the importance of particular sectors in an economy. The most common mistake is to confuse employment with output. The agricultural sector is insignificant in terms of numbers employed in the UK (about 2%), but is far more important in terms of output and contribution to the GNP, quite apart from its broader social and political importance. Similarly, the numbers employed in the automobile industry in the UK have declined sharply since the 1970s, but more cars were produced in the UK in 2000 than in any other year. This apparent contradiction is due to the fact that the automobile industry has replaced labour with machinery, thereby reducing employment but increasing output.

The influence of physical factors and the natural environment on industrial location

The natural environment and physical factors clearly have an impact on the choice of industrial location, although this varies greatly from industry to industry. Remember that 'industry' is just another word for work; it is a term that is interchangeable with economic activity covering any area of employment from primary to quaternary.

Physical factors

These involve small-scale site factors that are influential in the choice of location for individual plants or factories. Site factors include:
- availability of flat land
- ability to expand
- accessibility
- water supply
- ability to dispose of waste

In the case of some **basic manufacturing industries**, these factors have controlled the choice of site. An example of this is the development of estuarine sites for chemical industries (e.g. ICI at Billingham on Teesside).

At a more basic level, it is obvious that primary industries are strongly controlled by the location of the raw materials that are their reason for existence. For example, coal mining takes place where coal is found — an obvious physical factor. However, coal mining does not take place in all areas where coal is found. Economic, social and political factors determine whether or not it is actually mined. Thus, the decline of mining in south Wales is only partially explained by exhaustion of the resource, and much more by the price of imported fuel and the decline in demand for coal, as energy usage has been revolutionised in the last half of the twentieth century.

The natural environment
This is a broader term that might embrace anything from the general attractiveness of the landscape to the climate. It is of particular significance in those industries that are more flexible, i.e. industries that have low transport costs and can locate in a wide range of places with little impact on their total costs or their efficiency.

There has been much quaternary sector growth in areas perceived to be physically and climatically attractive, such as southern California and mediterranean Europe. An example of this is the location of Microsoft and Amazon in the American city of Seattle, because Seattle was a preferred choice of location for a large number of university graduates, long before the founding of Microsoft. The choice of location for some enterprises is driven by the need to attract the highest quality labour supply, which is relatively highly paid, geographically mobile and seeks the best possible environment in which to live and work. The persistent appeal of southern England, despite the higher costs of living, is attributed to a number of factors, but environmental appeal is one of these. At a more local scale, the development of a significant insurance sector in the southern English town of Bournemouth is partially explained by the attractive local landscape and the mild climate.

Weber's classical location theory and its weaknesses when applied to modern manufacturing industry
Location theory was developed by the German school of economic geographers, who attempted to build explanatory models and theories to explain the world. It was absorbed into geographical teaching at A-level in the 1970s, some 10 years after it made an impact in universities, along with a whole range of quantification techniques (essentially the use of mathematics and statistics) and tools of analysis. Before that, most locational work in geography had involved a detailed description of place and an attempt to recognise distinctive regions.

Models are simplified representations of reality, as shown in Figure 1.

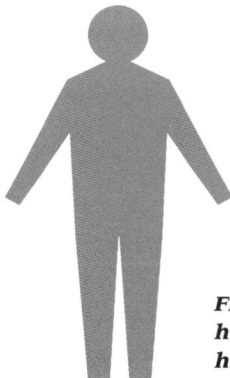

Figure 1 This is a model of a human being — it is obviously human, but omits much detail.

Models do not predict; they describe. Theories are predictive. They are based on a series of assumptions and try to isolate particular variables to make assertions that are testable. You will have used several models (such as the demographic transition model) and a few theories (such as those of Malthus and Boserup in population geography, concerning the relationship between population and resources) that are potentially testable.

Alfred Weber developed the dominant theory in industrial location. The theory predicts that industries will locate where they can minimise their **transport costs**. It is based on various assumptions, the most important of which are:
- inputs are available in unlimited supply
- demand is fixed and concentrated at a number of limited locations
- transport is possible in any direction
- transport costs are proportional to weight and distance

Weber also assumed that firms were independent enterprises producing a single product, and that choices made about location would be rational. He recognised that labour cost variations and possible savings to be made from co-location (agglomeration) might pull firms away from ideal cost-minimisation points. Much has changed since Weber wrote his book on location in 1909; he lived in a very different economic world from that of today, dominated by relatively small firms with one factory producing a single product.

The most obvious differences are that:
- there has been a revolution in transport that has much reduced the importance of transport costs as a proportion of the total cost of producing most manufactured goods
- there has been a fundamental change in the organisation and structure of industry, with the emergence of large corporations and global markets

Weber's theory and other related models have been criticised for their weaknesses over specific generalisations. These models can often be rescued by adapting them to a new set of conditions. For example, if transport costs are not directly proportional to distance, the model can still be used by making a few changes to the equal cost lines. However, the problem of the relevance of these models to modern industry is much more difficult to overcome. You should use your case studies to illustrate these difficulties. For example, the modern automobile industry has developed a quite different set of global locations from the modern textile industry. In the latter case, the price of labour has been the most significant driving force; whereas in the former, the quality of the labour force and agglomeration savings through external economies have been crucial.

Factors influencing present-day industrial location
This section is studied through the use of your case studies. You should have studied two contrasting manufacturing industries and will need to be able to draw on these to illustrate your answers.

Behaviouralist and structuralist explanations

Behaviouralist explanations are those whereby location is explained as much by social and cultural factors as by economic factors. Thus, many companies are situated where the original entrepreneurs themselves live (e.g. Dyson). These locations may be sub-optimal in strictly economic terms but may satisfy the owners and managers (hence satisfying motives). This can work at several levels; for example, the close relationship between universities (especially Stanford) and the development of Silicon Valley in California.

Structuralist explanations are those whereby location is explained by the underlying structures of society. Hence, as the economic system develops in particular ways, it requires particular types of location. Thus, location is driven by changes in the national and the world economy and, in recent years, the need to preserve profits in the face of a crisis.

The central theme here is that the factors that control the location of plants (factories) for a large **transnational corporation** like General Motors are not the same as those factors that explain the location of a smaller, owner-managed company. These differences in the structural organisation of companies will have a large influence on location. The textile and clothing industries illustrate some of the complexities of modern location.

Case study: the textile and clothing industry

- This comprises two industries: textiles and clothing.
- It officially employs 20 million people worldwide, plus another 5 million unofficially.
- It is the most geographically dispersed of all industries: and found in almost all countries. Why? It has low entry costs and is cheap to establish. The clothing industry especially is low-tech.
- It was the first industry in two distinct senses: historically, and in terms of development.
- The development of the textile industry in the UK set in motion the industrial revolution.
- Almost all economic development schemes had or have a role for the textile industry.
- It has undergone recent changes of location (as with the rapid rise of LEDCs such as China).
- LEDCs have 60% of the world export market (China alone has 29% of that total).
- Italy is the next largest (10%), but falling.
- The clothing industry is not automated (due to technical difficulties) and is thus highly labour-intensive.
- Wage rates worldwide are highly variable, for example 16 cents an hour in Indonesia and $7 an hour in Australia.
- There are two sources of cheap labour for clothing producers: citizens of LEDCs (usually women), or indigenous immigrant labour in MEDCs (mostly in large cities).

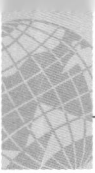

Case study: Australian outworkers

- There are 150 000 outworkers in Australia (usually recent Asian immigrants).
- Outworking is a response to the Asian challenge by Australian companies which find it hard to compete on cost and cheaper alternatives to hi-tech solutions involving expensive technology.
- Outworking takes place either in homes or in small sweatshops.
- It saves on expenses, such as the fixed costs of running factories, holiday pay, sick leave, pensions, etc.
- It employs mainly Asian women and their family members; there is no guarantee of further work, so there is a tendency to self-exploit, work through the night, etc.
- Asian immigrants often have very little alternative work and poor command of English when they arrive.
- Their general fear of government and authority leads to lack of any protests about conditions.

Case study: Nike in Vietnam

- Vietnam is one of the world's poorest countries, with a mean annual per capita income of $400.
- It has a population of 80 million, and 1 million new workers each year.
- There has been much unemployment in recent years as the centrally planned economy has slowly been dismantled by a more progressive government.
- As in many LEDCs, the clothing industry is one of the fastest growing sectors, employing 1.6 million people and earning $1.89 billion.
- Clothing firms are usually jointly run ventures with foreign firms based in Taiwan or Korea which have the expertise and markets, as well as the capital.
- Nike is the largest single private employer in the country, with 30 000 people employed by its agents.
- Its employees are mostly young women in their early 20s.
- In 1997, Nike was accused of exploiting labour and of failing to impose decent standards of health and safety.
- The campaign hurt Nike, which blamed local managers (who were often Korean).
- Nike overhauled its operation and raised pay to $47 per month; it now offers better working conditions than most other companies in the country.
- The campaign against Nike was backed by the anti-Communist movement in the USA which disliked the corporation propping up a corrupt regime in a country that provokes very bad memories for many Americans.
- A strike against Nike in 1997 led to major changes in its working practices.

The concept of 'made in the USA'

It is almost impossible to trace the origin of many products today. What exactly is meant by the label 'Made in the USA'? It is bought by patriotic Americans who prefer to support their own battered clothing industry by buying 'American'. The example of American Samoa introduces an anomaly.

American Samoa is a Pacific island that is an unincorporated territory of the USA. The largest factory on the islands is a clothing factory owned by Daewoosa, a Korean company, which has a largely Vietnamese workforce. Workers pay recruitment fees of $5000 because they are on US territory, and are employed to sew clothes with Sears, J.C.Penney and MV sports labels. Working conditions have been criticised and wages are thought to be well below Vietnamese levels.

As a result of this, you can buy a shirt in a US department store with a 'Made in the USA' label that was sewn by poorly paid Vietnamese women working for a Korean company on a Pacific island. This is one of the consequences of **globalisation**.

Employment structure, organisation and location are dynamic

Changes in the employment structure of the UK since 1945 and changes in the nature of employment

The main trends in UK employment since 1945 are:

- a continuing fall in primary employment
- an initial rise in secondary employment, followed by a long period of decline since the 1950s
- a particularly steep fall in secondary employment in the 1980s
- a rapid growth in tertiary employment, especially since the 1980s
- the emergence of a quaternary sector

It is vital that you do not confuse changes in employment with changes in output. The word 'decline' needs to be used with care here, since, for example, the automobile industry has certainly experienced a very large decline in terms of employment but not in output. In reality, output has been rising. Declining employment is due to the following causes:

- changes in demand pattern for some commodities (e.g. coal) and therefore changes in output and employment
- rationalisation and mechanisation (e.g. automobiles and electronics) replacing people with machines
- mergers of companies and economies of scale (e.g. automobiles), thus saving on jobs without sacrificing output
- profit crisis and plant closure (e.g. machine tools)
- plant closure as production has switched to cheaper, foreign locations (e.g. textiles and clothing)
- government policies and interest rates (e.g. ceramics and pottery)
- increasing affluence and more disposable income to be spent on services (e.g. the retail industry)

These changes confuse many students. The most common error is to assume that people choose tertiary employment because it is preferable to secondary employment. While this may be a minor factor, it is certainly not a major theme. Some of these changes are relative, in that a percentage fall in one category must lead to a percentage rise in other categories. It must also be remembered that total numbers in employment have risen by about 3 million since 1945.

The nature of employment has also changed. Some of the most significant aspects have been:

- a rise in female employment
- a rise in part-time employment
- a rise in temporary employment, with renewable contracts (often annually) and a fall in permanent employment
- the gradual polarisation of the workforce with many new jobs appearing in relatively poorly paid sectors and some in very well-paid jobs. The lost jobs of manufacturing were traditionally semi-skilled or skilled jobs which commanded relatively high wages.

Most of these trends can be explained by the need to cut costs and increase the flexibility of the workforce in order to maintain competitiveness. You might be able to illustrate them from your case-study industries.

The rise and decline of consumer industries in MEDCs

This section is concerned specifically with those industries that sell directly to the public — the **consumer industries**. These industries are almost entirely twentieth century in origin and are commonly associated with the rise of **Fordism**.

Fordism

Fordism involved the manufacturing of goods using assembly-line techniques, time and motion studies and mass production of standardised products. It also involved paying higher wages to workers who could, as a result, become consumers themselves. This was because for the first time, working people had high enough wages to buy the products that they made. This system expanded demand considerably, thus maintaining high profits for the companies through high turnover and economies of scale. Fordism takes its name from the Ford Motor Company, which was one of the first to establish this system.

Fordism dominated the world economy for the 50 years between the 1920s and the 1970s. During that time, Japan joined Europe and the USA as a major industrial power. Since the 1970s, many consumer industries have looked for new locations both within MEDCs and in new locations. They have done this in an attempt to reduce costs and, in some cases, to find new markets. Economists vary in their interpretation of this; some see it as the end of another Kondratieff cycle (see Figure 2).

The Kondratieff cycle

The **Kondratieff cycle** is a theory based on a study of nineteenth-century price behaviour. Data used included wages, interest rates, raw material prices, foreign trade and bank deposits. Kondratieff believed that his studies of economic, social and cultural life proved that a long-term system of economic behaviour existed and could be used for the purpose of predicting future economic developments.

Others regard crises of production as inevitable in the capitalist system, where there is too much factory capacity for the demand. For example, the rapid growth of postwar economies involved rapid rises in demand for goods as the Western world stocked up with refrigerators, hi-fi systems and, above all, cars. Inevitably, demand slowed down when most of the people who could afford such things had acquired them. The factories that had been turning out goods to meet a rapidly growing demand from new customers were now faced with largely 'replacement' customers who simply sought to change their old models. This led to the idea of overcapacity and hence the crisis. Your case study industries should provide you with some details about such changes.

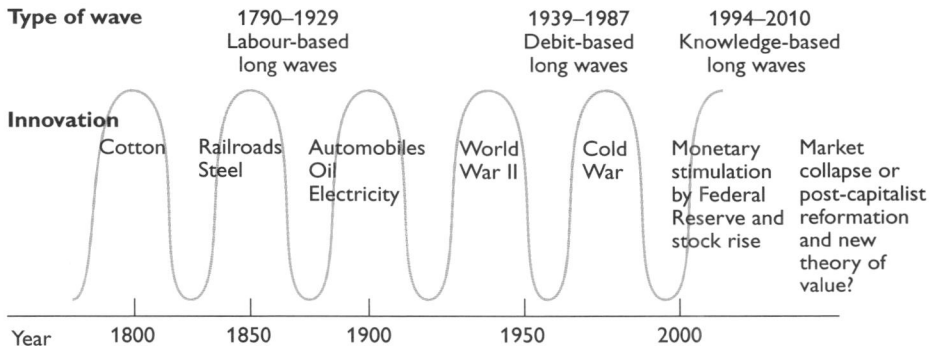

| Type of wave | 1790–1929
Labour-based
long waves | | | 1939–1987
Debit-based
long waves | | 1994–2010
Knowledge-based
long waves | |

Innovation

| Cotton | Railroads
Steel | Automobiles
Oil
Electricity | World
War II | Cold
War | Monetary
stimulation
by Federal
Reserve and
stock rise | Market
collapse or
post-capitalist
reformation
and new
theory of
value? |

| Year | 1800 | 1850 | 1900 | 1950 | 2000 |

Figure 2 The Kondratieff cycle

Common mistakes include a tendency to generalise too much about MEDCs. The speed of change has been much more noticeable in some than in others. The decline in consumer industries has been limited in Germany and sharp in the UK, while in Spain, there have been recent increases in manufacturing employment. This has also been the case in other parts of mediterranean Europe and in parts of the old Soviet empire.

The rise of manufacturing in NICs and the environmental impact of that growth

There is no official list of newly industrialised countries (NICs) and no agreed definition of them, but they commonly include the main four which are known as the Asian 'tigers':

- South Korea
- Taiwan
- Hong Kong
- Singapore

There are also a number of 'tiger cubs', including:

- Malaysia
- Thailand
- Indonesia
- Philippines

In addition, there are a number of more complex and speculative cases, including:

- India
- Brazil
- China
- Mexico

Many of these countries, especially the Asian tigers, have experienced very rapid growth in gross national product (up to 10% per annum sustained over several years). They have achieved this by developing a manufacturing sector. This contrasts with the equally rapid growth experienced by countries like Uruguay earlier in the twentieth century, or many oil states today, which grew by exporting primary products.

NICs are not a uniform group and should not be treated as such. They represent success stories in spreading the benefits of economic growth to parts of the world outside the core countries. For much of the last century, industrialisation was restricted to North America and Europe. The rest of the world was a peripheral area, much of which was utilised by the core areas to provide raw materials through one system or another (e.g. through colonialism, neo-colonialism and dependency — see pages 57–59). NICs broke away from this arrangement and you should be aware of:

- how they did it
- who helped them
- the problems caused by this

The answers will depend on your choice of country or countries to illustrate this, but look out for:

- export growth behind tariff walls
- the role of transnationals and foreign investment
- the environmental costs

One point to remember is that two of the NICs (Hong Kong and Singapore) are rather odd examples because they are really city-states and have no territory other than the city itself. Therefore, they are trading centres that are not easily comparable with countries like South Korea and Taiwan, both of which have had very considerable American influence in the last 60 years.

The emergence of a new international division of labour

The new division of labour is a result of a range of trends. For example:

- quaternary services, especially research and development, are found in the core world regions of Japan, Europe and the USA
- manufacturing activity is increasingly found in the NICs, where labour costs are frequently much cheaper
- primary production is still dominated by the peripheral countries, but is complicated by the fact that, for example, the USA is the world's major exporter of wheat
- the new information-led economy works on a global scale, but markets are far from fully integrated, and governments and regional trading blocs (e.g. the EU) still play crucial regulatory roles
- there are regional differences within the three areas of influence: North America, the European Union and the Asian Pacific region (dominated by Japan)
- Around this triangle, the rest of the world is organised in a hierarchical and inter-dependent web, as different countries and regions compete to attract capital, human skills and technology.

The global capitalist economy is also divided according to dominant technology:

- producers of high-value goods based on 'knowledge' and skills (quaternary and hi-tech)
- producers of high-volume goods, based on lower-cost labour, such as trainers and T-shirts
- producers of commodities such as coffee and cotton
- producers of material based on natural resources, such as diamonds and timber
- producers who remain outside the global capitalist system, such as subsistence farmers in Somalia

This division does not coincide with particular countries; all countries have this type of division, but the distribution varies greatly.

Two other factors that should be mentioned are:

- the emerging significance of global cities which are often the control centres of this system of production
- the central role of the transnational corporations in both the development of this system and its further refinement

Remember that there are major generalisations involved in this sort of analysis. You should already be familiar with some of them, including:

- the local concentration of the quaternary sector (not all of Europe has been favoured by the growth of these industries)
- the existence of pockets of cheap labour and sweatshop-style industry in global cities (e.g. outworkers in Australia)
- the patchy decline of manufacturing in the core world regions and the growth of manufacturing in Spain

The globalisation of industry and the role of governments are increasing

Transnationals, the globalisation of production and the concept of glocalisation
Transnational corporations (TNCs) are:

- large
- managed by professionals who do not own the company
- operate in several different countries

TNCs have been around for more than a hundred years, but their recent growth has been considerable. Although they may bear the name of a country on their logos (e.g. British Petroleum or BP) and have their headquarters in their country of origin, they owe no particular loyalty to those countries. They are driven by the need to maintain profitability to reward shareholders; therefore, they are constantly searching for more efficient methods of production and cheaper locations for that production. 500 TNCs produce 30% of planetary output and account for 70% of trade. Some consider that they are so powerful that they become more important than governments.

Globalisation is the process whereby the whole planet has been incorporated into the capitalist system of production. It involves an increasingly free flow of capital (money and machinery), reduction of tariffs and the growth of free trade. It is not entirely new; at the height of the British Empire, a similar system prevailed.

Glocalisation is a system of production pioneered by Japanese automobile producers whereby production takes place outside the home country but, in contrast to production in assembly branch plants, local (hence **glocal**isation) sourcing of parts is involved. This allows producers to minimise the holding of stocks of parts with virtually daily delivery from local suppliers. This is part of the JIT (just-in-time) production system that helps minimise costs.

Opponents of globalisation point out that of the 200 largest TNCs in the world, 172 have their world headquarters in the US, Germany, UK, France or Japan. They argue that these headquarters remain there because these TNCs benefit from close ties to their own national governments. Such benefits include tax breaks, favours from political leaders, links to education and training institutions, use of infrastructure, and the possibility of using the state military or police in times of crisis.

Opponents of globalisation also argue that research and development departments of TNCs typically remain in the home-base country. This is because, in their view, the knowledge and skills of new technology are central to the power of TNCs and certain states. If research and development were spread throughout the world, people in all countries would develop the capacity to make the latest computers or cell phones and would then be able to compete in the world market. For this reason, research and development is a carefully guarded secret within most TNCs. For example, it is alleged that only a handful of people know the actual chemical formula and process for making Coca-Cola or the recipe for Kentucky Fried Chicken.

You need to have studied at least one major TNC and have some details about its locational decisions and the history of its development. Use examples to illustrate the increasingly global nature of production and some of the issues outlined. This is a highly controversial topic; there is increasing opposition to the apparent dominance of TNCs on the world stage, to the policies that are driven by them and to those governments which seem to collaborate in the growth of this world system.

The impact of globalisation on employment structures throughout the world
Employment structures have obviously changed as a result of globalisation. The key aspects that you need to be able to expand on are:

- the **growth of secondary employment** in some countries and regions (e.g. the NICs)
- the **decline of secondary employment** in some core countries (e.g. the UK experience)
- the **changing pattern of primary employment** in some peripheral countries (and the increasing importance of commercial agriculture for the production of export crops)
- the **growth of quaternary employment** in key regions of core countries

There are both positive and negative impacts to be noted; you can do this on both a local and national level. A particularly good source of detailed information about these processes is the changing pattern of employment in Mexico and the USA over the past few years as a consequence of the formation of NAFTA.

Case study: The maquiladoras

US-owned **maquiladoras** are assembly plants in Mexico that employ Mexican labour to make products for export to the US. In 1965, the Mexican government set up the Border Industrialisation Programme which created export platforms for US companies on favourable terms. The word *maquiladora* is derived from *maquilar*, which means 'to submit something to the action of a machine'. When US companies opened factories in Mexico, the name *maquilar* evolved to refer to the process of assembling parts manufactured elsewhere.

- Approximately 2100 *maquiladora* plants produce electronic goods, automobile parts, chemicals, furniture, machinery and other goods in Mexico. The number has increased more than fourfold since the mid-1980s. About 600 000 workers are employed in the *maquiladoras*. Ninety per cent of the plants are US-owned.
- Many of the largest US corporations have *maquiladora* plants. These include AT&T, Ford, General Electric, General Motors, Dupont, Eastman Kodak, Emerson Electric, IBM, ITT, PepsiCo, United Technologies, Xerox and many others.
- *Maquiladoras* receive government subsidies such as preferential tariffs and taxation. They pay no tariffs on materials and semi-finished products imported into Mexico. When they ship finished products back to the US, they pay tariffs only on the value added in Mexico, not the value of the entire product.
- *Maquiladoras* have continued to grow and develop since the formation of NAFTA in 1994. The impact on employment in US manufacturing has been significant, as has the growth of manufacturing employment in Mexico.

Positive factors include the very rapid economic development of northern Mexico and the benefits of that development, but there have also been negative environmental impacts as a result of this growth. Other negative factors include the impact on the distribution of income within the country which, in some periods, has become more unequal, with potentially damaging effects on social and political stability.

Political processes influencing the location of industry

The term 'political processes' does not only involve governments; it extends beyond governments to include the role of the many NGOs (non-government organisations) that have an increasingly powerful influence on political decisions. It also applies to the local scale of planning controls and the development of the infrastructure at a local and national level.

You should be careful to distinguish between **direct** and **indirect** influence. The use of regional policy with incentives and tax breaks on offer is obviously direct government action, whereas the building of a new motorway or bridge is indirect, in that it is not targeted exclusively at industry.

There are several scales here, from the *maquiladoras* example, in which government

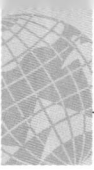

action on both sides of the border has been central, to a very local level of recent protests about the location of a chemical plant in Gloucestershire, which has led to a fierce debate about safety between local residents and local government.

It is quite legitimate to discuss any type of industry as an example, so the relocation of offices from London or the development of Cardiff Bay would provide material for showing the impact on location.

The emergence of large trading blocs and their impact

These are part of the wider issue of globalisation. Their emergence generally involves an effort to remove barriers to movements of capital, goods and labour. At a simple level, this creates free trade associations, for example Mercosur; at the most complex level, it creates an overarching trend towards full political integration, for example the European Union.

The origins of large trading blocs lie in the post-war economy, although the imperial system of free trade was an obvious precursor. The original super-state was the US. The US has theoretical advantages, which are rooted firmly in classical economic doctrine laced with neo-liberalism. The theories of comparative advantage and economies of scale provide the scaffolding.

The **law of comparative advantage** states that countries should specialise in producing and exporting the goods that they produce at a *lower relative cost* than other countries. Where a country is more efficient in the production of two commodities than another country, it should specialise in the commodity in which its comparative advantage is greater.

The principle of **economies of scale** states that the greater the scale of production, the lower are the average costs of producing each unit. The ideology behind this is the presumption that if all 'barriers' are removed, production will become more efficient, bringing benefits to all.

Practical policies *always* include:
- removal of tariffs
- establishment of trade rules, such as anti-dumping procedures
- deregulation of services, for example the US frequently considers state services as a subsidy to employers and therefore unfair
- national treatment of foreign investment, removing local content provisions whereby a stated proportion of parts has to be produced in the 'host' nation
- intellectual property rights, ensuring copyright regulations are strictly enforced

Practical policies *sometimes* include:
- removal of passport controls
- common currency
- integration of political and legal systems

Supporters argue that to achieve full economic integration, political and social integration must also be achieved.

Rural–urban interrelationships

The process of urbanisation is dynamic and varied

The location and distribution of the world's major urban areas

Urbanisation is the process by which there is an increase in the proportion of people living in urban areas. Common methods used to define towns are: population size; population density; function; and level of administration.

The United Nations (UN) does not classify settlements as towns or rural areas, but has instead chosen to classify them by size. This varies according to the census methods of individual countries, making the figures for city size unreliable in detail, although the big picture emerges clearly enough.

The following tables show the distribution of large cities by region (1800–2000).

Number of million-plus cities				
Region	**1800**	**1900**	**1950**	**2000**
Africa	0	0	2	34
Asia	1	3	26	136
Europe	1	9	30	61
Latin America	0	0	7	39
North America	0	4	14	36
Oceania	0	0	2	5
Total	**2**	**16**	**81**	**311**

Number of the world's largest 100 cities				
Region	**1800**	**1900**	**1950**	**2000**
Africa	4	2	3	6
Asia	64	23	32	44
Europe	29	52	37	19
Latin America	3	5	8	16
North America	0	16	18	13
Oceania	0	2	2	2

Average size of the world's 100 largest cities				
Region	**1800**	**1900**	**1950**	**2000**
Population	187 000	724 000	2.1 million	6.2 million

These tables show that the growth of large urban areas (cities) has been especially notable over the last 50 years. The distribution has changed significantly from the dominance of Europe (in 1900), being first challenged and then overtaken by Asia, which has a long urban history — see 1800. Africa has also emerged as a continent of rapid urbanisation in the last few decades and, significantly, the continent that remains the least urban. You should be able to describe this pattern of global urban growth.

Reasons for, and factors affecting, urbanisation as a process

Figure 3 shows the proportion of world population living in urban and rural areas.

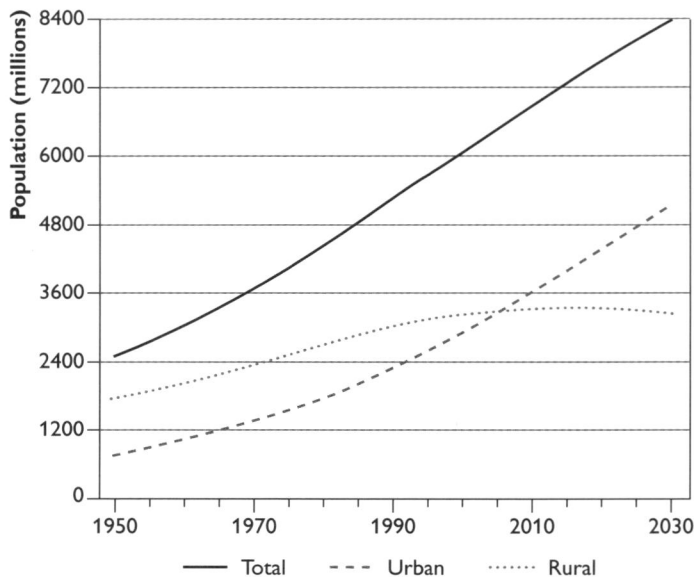

Figure 3 Projected total, urban and rural population

Chronology from pre-industrial to post-industrial cities

This can be outlined as follows:

- **Pre-industrial cities** are those urban settlements that existed before the early nineteenth century or those in countries that have yet to show many signs of transition to industrialised economies. They are referred to as pre-industrial in the sense that they developed without a core of industry acting as a centripetal force attracting people to one location. Their functions varied, but were dominated by administration, government, religion and trade. They were often military strongholds and centres of authority. All of them had some industry in the sense that things were made there, but it was neither the primary function of these cities nor the cause of their development. In that sense, manufacturing was not city-forming. It was city-serving, in that the inhabitants needed clothing and household articles and an industry developed to serve these needs.

- **Industrial cities** developed as a consequence as the growth of manufacturing industries, which drew people together at one site as the labour force for those industries. Nineteenth-century city growth in Europe and North America was largely based around the development of new industries — initially heavy (capital) industries and then consumer industries.
- **Post-industrial cities** have developed more recently as the location of industry has shifted away from traditional centres and become less concentrated. Some cities have developed around different types of economic activity, especially tertiary and quaternary services. More commonly, older cities have found non-manufacturing employment of growing significance and have lost older city-forming activities such as their traditional manufacturing core. These cities are characterised by low densities and, often, a high dependence on the automobile.

These definitions are essentially chronological, but you should note that urbanisation without industrialisation has been a significant theme in the last 50 years, especially in Asia and Africa. Thus, many of the **new million** cities are not copies of the cities that grew in nineteenth-century Europe and North America.

Processes of urbanisation

At the demographic level, urbanisation involves two processes:

- natural increase
- migration

Rates of **natural increase** in cities are often significantly higher than those in more rural areas, with higher fertility reflecting lower age profiles, but the differences in age-structure are themselves the consequence of migration. In many cases, it is relatively younger people who arrive in cities.

Rural–urban **migration** has been the dominant force in the development of cities. Remember that there are likely to be factors operating at both ends of this process; in other words, 'push' factors from rural areas and 'pull' factors from urban areas. Of the one thousand or so migrants who arrive each day in Mexico City, the majority have no specific and fixed employment to go to. This creates the common stereotype of urbanisation without industrialisation.

Your own case studies should allow you to fill out this picture and above all to qualify some of these generalisations. The following points should be noted.

- Urban growth based on industry pre-dates the Industrial Revolution. Potosi in modern-day Bolivia was a major city based entirely on the primary industry of silver mining.
- Manufacturing employment exists in all cities, but is not the primary cause of development of all cities.
- There is a large informal sector in many LEDC cities of manufacturing employment that takes place in sweatshops or in homes.
- Some post-industrial cities such as Los Angeles have a significant manufacturing sector, although much of it is on the physical margins of the urban area.

- Urban growth in modern Africa is more due to changes in rural areas than to industrial growth in urban areas.
- The causes and rates of urban growth are variable within continental areas. Thus, the spectacular growth of Shanghai is due to growth in secondary employment, but the growth of Hong Kong today owes much more to trade and commerce.

Global variations in the rate and characteristics of present-day urbanisation

Cities come and go. The forces that create and destroy cities today are generally global and dominated by the same economic forces outlined in the Economic Systems section of this module. However, some cities are remarkably resilient — whether it is London, Damascus or Beijing, it is obvious that some city-forming forces are very powerful.

In this section, you will need to be aware of the various types of modern-day urban change. At least five categories can be recognised.

- The old industrial cities that have lost population in the last 30 years as their traditional industrial base has declined. Many of these are in Europe or North America, for example Sheffield, UK and Cleveland, USA.
- Old industrial cities that have had some success in replacing their traditional manufacturing base with new types of employment, thus maintaining population or experiencing some growth, for example Milan, Italy and Boston, USA.
- Rapidly growing cities based around new technologies and/or tertiary and quaternary industry. These are the post-industrial cities, for example Houston, USA and Toulouse, France.
- Rapidly growing cities in NICs (newly industrialised countries) that are major centres of administration, education and government as well being manufacturing cities, for example Seoul, South Korea or Tai Pei, Taiwan.
- Rapidly growing cities in LEDCs that are subject to considerable levels of rural–urban migration driven, principally, by rural poverty and the commercialisation of agriculture. The landless poor become the urban poor — a situation found in almost any major African city, for example Khartoum, Sudan.

You need to be aware that any such categorisation is very tentative; some cities resolutely resist simple classification. Many of the **global cities** referred to in the previous section have been remarkably adaptable and their success has been due to speed of adjustment to global changes. London has been at various times, and in various combinations, a major manufacturing city, a major trading centre, a financial capital, a centre of government and a cultural capital of world significance. It has experienced population loss but, as with New York, the counter-urbanisation process involved is the spreading of the influence of the city rather than a sign of decline.

The economic and political processes in the management of cities

The management of cities has become a significant issue over the last hundred years. During the rapid growth phase of industrial cities in the nineteenth century, cities were not, by and large, managed at all. Hence, Birmingham and Manchester grew in an unrestrained and uncontrolled manner. Manchester had a population of 160 000 in 1832 but no government other than a medieval manorial court! This led to problems that were addressed initially by wealthy industrialists who saw that the impact of disease and unsanitary conditions threatened the whole process of industrialisation, as their workforce became less and less productive.

The current global situation is that:
- many world cities are facing crisis, especially the most rapidly growing urban areas in Africa and Asia
- up to 60% of the urban inhabitants of African cities live below the poverty line
- 40% of the global urban population have no access to safe drinking water

The results of this are obvious enough, with very high infant mortality being a particularly grim characteristic of large urban areas in much of the world.

United Nations Conference on Human Settlement, 1996

The stated aims of urban management at the United Nations Conference on Human Settlement, held in Istanbul in 1996, were 'to make cities healthy, safe, equitable and sustainable'. The conference recognised three critical areas that need to be solved:

- **Housing**. International law recognises basic housing as a human right. Up to 500 million urban inhabitants have no homes and the same number live in conditions that are insanitary and dangerous. Your own case studies should allow you to illustrate a range of management policies used to address this problem.
- **Clean water, sanitation and waste management**. The most important causes of death and disease are those related to lack of clean water and an inadequate sewage disposal system.
- **Public transport**. In almost all of the world's large cities, public transport systems have been under considerable pressure. High levels of private vehicle usage have led to serious deterioration in air quality, and journey times have scarcely improved in many cities.

There are other management issues that could be addressed in this section. An increasing number of cities plan for growth to maintain competitiveness in the global economy by attempting to provide the infrastructure required by international capital. In other words, how do you make a city attractive to foreign and domestic investors? This may include economic development, such as transport facilities (e.g. Birmingham) or cultural facilities (e.g. Barcelona and Glasgow).

You should be familiar with the idea of **sustainable development**, which embraces not only the natural environment but also social and economic goals. For example, public housing provision through subsidies has provided a base for economic development in Hong Kong, whilst in Singapore, the application of ethnic ratios has helped avoid high levels of racial and ethnic segregation, which are so common and so divisive in other cities.

In most rural areas, agriculture remains the dominant land-use

A classification and description of agricultural systems at a global scale

Subsistence farming

Subsistence farming systems aim to grow only enough food and fibre for their own needs, collect fuel and building materials from natural sources, and hardly enter into the cash economy. About 30% of the world's population and about 60% of the world's farmers are living on small subsistence farms. Many African countries and much of southern and eastern Asia contain a high proportion of intensive subsistence agricultural systems of which the most important for human food supply is wet-rice agriculture.

Commercial farming

Commercial farming systems produce agricultural commodities for sale. Capital is used to purchase items, such as tractors and machinery, fertilisers, pesticides, improved plants, better breeds of animal, and other technological innovations. These farms tend to be larger than subsistence units, but very small farms (even less than one hectare, such as some of the commercial rice farms of Japan) are also classified as commercial. The large amount of capital required, including money, equipment, labour and management, tends to restrict ownership and favour operators of large-scale units. This type of farming system includes:

- tropical and sub-tropical plantations, such as LEDC banana and sugar cane growers
- mid-latitude grain farming, such as North American and European wheat farmers
- vegetable and fruit cultivation
- mixed crop and livestock farming
- livestock ranching

Collective farming

Collective farming systems are usually located in countries with a centrally planned economy, such as mainland China, North Korea, Vietnam and Cuba. In these societies, agricultural production operates under a system of collective and state farms. Land is collected into a single unit of operation under government supervision. The workers tend the land and receive a small salary, but do not receive any credit or profit for anything produced. Some private ownership is allowed. In some countries, small private plots are provided for members of collective farms, state farm employees and other workers. These systems are generally not as productive as comparable privately owned operations.

It is important to remember that all of these systems can co-exist in the same local and national areas. There is no doubt that commercial agriculture has grown at the expense of subsistence farming over recent years, but there has been much argument about their relative efficiencies. It used to be suggested that commercial agriculture was necessarily more efficient than subsistence farming but, in reality, a more complex picture emerges. It depends a great deal on the criteria used to measure efficiency.

Intensity of agricultural systems

You also need to be able to differentiate between levels of **intensity** of agricultural systems.

- Intensive farming uses high levels of inputs (labour and/or capital) per unit area of land. It is characterised by high output per unit area.
- Extensive farming uses low levels of inputs per unit area of land and achieves low outputs per unit area.

A common error is to assume that intensive agriculture is more profitable than extensive agriculture and hence, in some way, preferable. Of course, it is more profitable per unit area, but agricultural systems that tend to be extensive (e.g. cattle-ranching) may involve very large areas of land per farm and can be just as profitable as smaller, more intensive, farms.

Pastoral and arable systems

A further categorical distinction can be made between **pastoral** and **arable systems**. In almost all cases, arable farming gives higher returns of food per unit area and is thus more efficient. Pastoral farming is often a response to a constraining factor within the environment, which disallows arable farming.

The global land area is roughly divided into thirds:
- 35% is used to produce agricultural products
- 31% is used to produce wood and timber
- 34% is in non-agricultural usage

Within the 35% used for agricultural products:
- 24% is pasture and meadowland used for animal production
- 10% is arable used for annual crops such as grains and cereals
- 1% is used for permanent crops such as some fruit and nut trees

A total of 3.1 billion hectares of land throughout the world is potentially arable. Of that, only about 1.5 billion hectares are being used for field and horticultural crops. Much of the unused potential land is found in areas that lack essential transport or other infrastructure needed for the maintenance of commercial food production.

Virtually all human nutrients come from three sources: crops, animal products (meat, milk and eggs) and aquatic foods. About 3000 plant species have been used for food and more than 300 are widely grown; however, only 12 provide 90% of the world's food. Crops furnish over 90% of the human population's calories and about two-thirds of its protein.

Among crops, the cereals (e.g. wheat, rice, corn/maize and millet) occupy three-quarters of the crop area and provide about three-quarters of the world's calories. Root crops, oilseeds and sugar provide about 20% of the calories from crops. Vegetables and fruits satisfy only a very small proportion of caloric needs, but are very important for other nutritional values.

The physical constraints of agricultural systems

The location of agriculture and the development of particular agricultural systems are inevitably influenced by the characteristics of both the physical environment and the human environment. The type of agriculture and whether it is commercial or subsistence will determine which factors are the most influential in determining location. The degree to which physical factors can be used to explain variation depends on the scale. On a global scale, there is no question that climate is the dominant influence, especially precipitation and the length of the growing season; but on a more local scale, soils and relief explain variation better.

Physical factors
- Soil
- Climate
  - Precipitation
  - Length of thermal growing season
- Relief
- Aspect

You need to be able to illustrate the limitation imposed by the physical environment with the use of detailed examples drawn from both MEDCs and LEDCs, from both commercial and subsistence agricultural systems. One possible approach to accounting for the physical constraints on a global scale would be to explain the pattern shown in Figure 4.

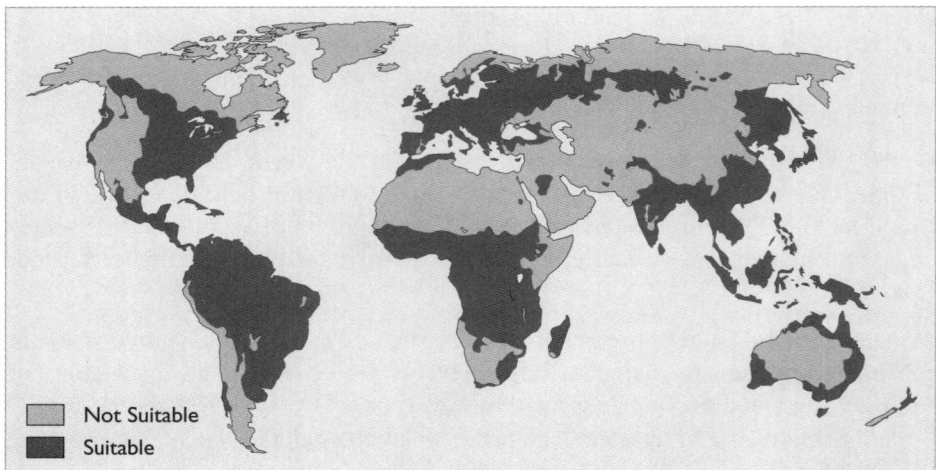

Not Suitable
Suitable

Figure 4 Areas where climate is suitable for crop agriculture

This should involve the use of other world maps of precipitation and temperature to build a picture. At a different scale, variation of cropping patterns within a country might be explored. At this scale, other physical factors can be introduced to explain variation. These are likely to include soil quality and relief factors, as well as availability of water for irrigation in the case of many arable crops, including cotton (Figure 5).

Per cent area
☐ 0
▨ 1–6
▨ 7–16
▨ 17–29
▧ 30–37
■ >37

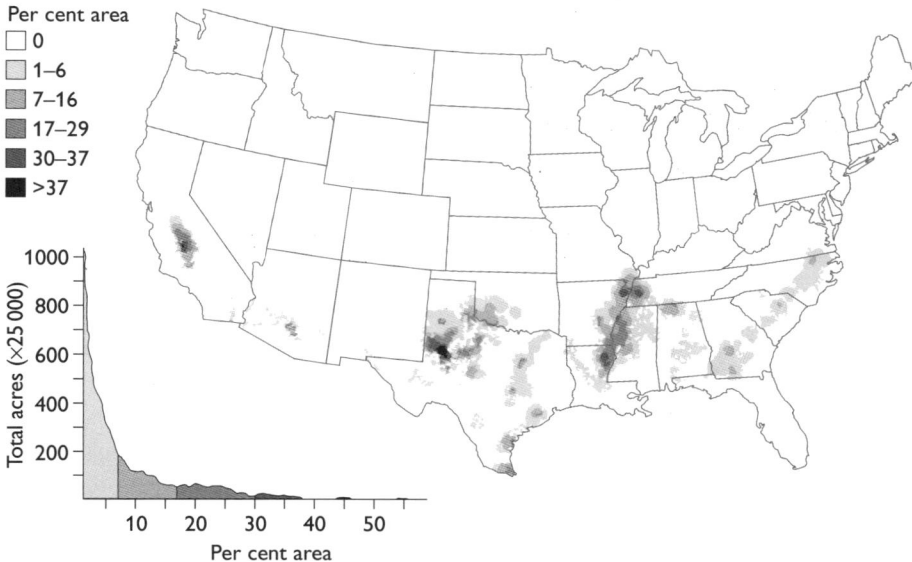

Figure 5 Distribution of US cotton, 1992

Social, economic and political factors influencing agricultural land use

Having established the physical controls and constraints, it becomes obvious that a great deal is left to explain. The simplest approach is to change scale yet again and look carefully at variations in agricultural land use within the local area. In the UK, as elsewhere, this is dominated by human factors such as systems of land tenure, inheritance laws, accessibility, markets, competition and government.

Systems of land tenure

These range from large estates held often for tax purposes (uncommon in modern Europe, but important in Latin America) through to sharecropping and plantation systems. In Europe, the key distinction is between the farmer as owner, the farmer as tenant and the farmer as manager. The impact on agricultural land use is often determined by the different access of these various groups to capital for investment.

Inheritance laws

In general terms, farm size has increased — very sharply so in some world regions, such as Europe. This is often associated with increasing levels of commercialisation in agriculture. In some areas (much of continental Europe), the fragmentation of land on the death of the owner is a consequence of inheritance laws that compel land to be left to all surviving children equally. Dowry customs in many LEDCs have also led to fragmentation. These social factors have been seen as negative influences in the development of commercial farming.

Accessibility

Transport availability and distance from markets impact on price; inaccessibility impacts both on the possibility of preserving agricultural products and on the cost of transport.

Even in a world where transport costs have fallen very sharply, this has a continued influence. The distribution and type of dairy production is an obvious example.

Markets

The size and type of markets are very significant. Differing and changing tastes in the face of competing cultures (e.g. hamburgers in Kenya) clearly have a profound influence. Intolerance of dairy products in some ethnic groups is an obvious example. For instance, there is a lack of large-scale milk production in China, even in areas where it is climatically suitable; whereas the existence of a dairy industry in the sub-tropical southern USA is an indication of the popularity of milk products among Anglo-Saxons. Increasing affluence has made a considerable impact on demand patterns in the UK; the rising popularity of organic products represents a considerable shift in emphasis that is highly significant for producers.

Competition

The increasing globalisation of food production has meant that farmers have had to face increasing competition from imported products; the trend to reduce tariff barriers has emphasised this. In the past, a country such as Uruguay specialised in the production of beef and beef products for export. It lost its market in the 1950s as other producers developed their own beef industries and the original market (mostly European) built itself a new agricultural policy in the post-war period. The impact on Uruguay was catastrophic.

Governments

This is by far the most significant of these variables and needs close attention. Governments have interfered in food production for a very long time, but during the twentieth century the extent and impact of government interference have increased greatly. The most productive route is to look at the changes in European Union policies towards agriculture over the last 50 years and the impact of these policies on food production and patterns of cropping. The important social changes that have taken place should also be addressed, including the decline of family farms and the rise of agribusiness. The increasing role of the World Trade Organisation (WTO) and the TNCs in developing global markets and global production patterns challenges some of the power of national governments in controlling their own agriculture.

The globalisation of food production and the politics of that global production

This is not a new theme. During the seventeenth century, the European 'North' created a colonial system based on plantation agriculture in the 'South', with these colonies being deliberately and specifically designed to produce agricultural products for export to the home country. After decolonisation, a complicated and varied set of policies was pursued, sometimes with the aim of increasing the independence of the former colonies, but often in reality making them even more dependent on the richer 'North'. For example, despite efforts (including a revolution) to break out of this system, modern Cuba is still heavily reliant on the production of sugar, first introduced there in the seventeenth century by Europeans using slaves for labour.

Globalisation of food production is obvious, at a glance, when walking around a supermarket. Vegetables from West Africa, fruit from New Zealand, wine from Argentina and tea from India compete for space on the shelves. The causes of such developments are:

- improvements in the technology of transporting perishable products
- promotion by large retailers and giant food companies of new products
- standardisation of food production techniques
- reduction in tariffs and other barriers to trade

These processes need the consent of governments (even if given reluctantly) — hence the central role of politics in driving this globalisation of food production. At the heart of this debate lies a paradox. Food is not in short supply. In truth, food products have never been so plentiful; globally, we produce enough to feed each human being 2700 calories a day. However, about 30 million die of starvation every year and a further 800 million suffer from chronic malnutrition. Some of these people live in the very countries that are increasingly committed to exporting food products to stock the supermarket shelves in the richer MEDCs. Thus, famine and malnourishment are consequences of the inability of people to afford to buy food rather than the unavailability of food. This is a reminder of the population and resources debate that was begun by Malthus and still continues today.

The global forces that drive these changes are very powerful and are certainly not restricted to the production of food products. The impact of the globalisation of food includes:

- increasing variety of products in MEDCs
- increasing areas devoted to the growth of export crops in LEDCs
- increasing control of food production by large corporations
- increasing development of chains of production from agribusiness through to major retailers

More controversially, it also includes:

- increasing dependence of LEDCs on MEDCs
- increasing rural poverty as poor peasant farmers lose their land
- increasing urban malnourishment as LEDCs switch land used for domestic food production to export crops
- changes in consumer tastes in LEDCs, introducing new products to replace traditional diets

A related issue in the politics of food production is the fierce debate about the role and the control of technology in agriculture. The agricultural revolution predicted by some, with the introduction of genetically modified seeds in the USA, has run into difficulties, as European consumers have started to show some concern about the possible health implications of these new crops and the manner of their promotion by the large food corporations. Governments, through their responsibility for the nation's health, have had to develop policies to deal with this issue and other health-related concerns.

There is an interdependence between urban and rural environments

The changing resource base of the rural environment, especially in MEDCs

It is obvious that rural areas are no longer as distinctive as they once were. In pre-capitalist societies, rural could be defined as:

- areas dominated by the production of food, both in terms of employment and area of land
- areas dominated by settlement forms that were largely villages, hamlets or isolated farms

The role of agriculture has declined in two senses:

- less land is devoted to farming as other uses have developed
- employment in agriculture has declined very sharply, reducing the economic significance of this sector within local rural communities

As a result, the nature of both rural environments and their resource base have changed.

Amenity areas

These rural changes include the growth of amenity areas. Amenity areas are an ancient concept. The New Forest in Hampshire was once a royal hunting area and its landscape of mixed forest and heathland still bears the imprint of this. The use of rural areas as an amenity has increased rapidly in the twentieth century. The first National Parks were created in the USA; they are 'national' in the sense that they are owned by the nation as a whole. These are wildernesses from which all human permanent activity is excluded. This is not the case in the UK; the National Park in the Lake District, for example, is not owned by the nation but largely made up of areas of privately owned land. Nor is it really a park. It is a landscape much altered and affected by man, in which human occupation is visible and permanent, but which is regarded as exceptionally attractive. These areas are preserved by stricter planning controls than apply elsewhere.

In these regions and many others rural areas, leisure activities have made an impact. These include:

- walking
- climbing
- water-sports
- golf
- birdwatching

New resources have appeared, including:

- theme parks
- heritage trails
- country parks

A whole range of accommodation and other facilities has also arisen in response to growing numbers of day-trippers as well as short-stay or long-stay visitors.

These include:
- hotels and motels
- bed-and-breakfast accommodation
- camp sites
- cafés and restaurants
- souvenir shops, gift shops and antique shops

You should note that the growth in these amenities has been highly localised and often in marked contrast to the decline in more traditional rural functions, such as the village shop and the pub.

There have been other pressures on rural areas, including the extraction of minerals and the development of reservoirs. Although significant locally, they have not had the same impact as the growth of leisure activities and tourism.

The other major impact on the rural resource base has been the further blurring of the distinction between urban and rural areas as industry other than agriculture has developed in rural regions. This is dealt with in the following section.

The patterns apparent on rural/urban fringes in MEDCs and LEDCs, and the processes involved

Three categories of urban growth on rural/urban fringes are explored in this section, all with contrasting processes:
- cities in MEDCs (especially in the USA and Europe)
- cities in NICs (especially in southeast Asia)
- cities in poorer LEDCs

Cities in MEDCs

The physical growth of cities, the impact of commuting and the decline of inner-city populations have had a profound impact on the geography of the rural/urban fringe (i.e. those areas on the edge of the built-up area of a city). There are distinctive differences among MEDCs. In Japan, the density of city populations and the intensity of agricultural land use on the margin have severely restricted urban growth. At the other extreme, in the USA, this boundary is now so blurred that two new urban forms have been identified:
- edge cities
- post-suburban regions

According to most definitions, areas become **edge cities** when they have become urbanised within the last 30 years, and have at least 5 million square feet of leased office space and 600 000 square feet of retail space, along with a population that increases at 9 a.m. on workdays.

Post-suburban regions either have no specific centre or are multi-centred. Such regions are not just multi-centred in their commercial activities. Their commerce, shopping, arts, residential life and religious activities are all conducted in different places on a network of interconnected travel paths linked primarily by private cars.

The quote below from a resident of Fairfax County in Virginia is typical of the dependency on cars in many US households:

> My family has four cars. I drive only one, but the other three don't sit idle. My wife, my daughter in college and my daughter in high school each use a car every day. There's not much chance I'll go back to two or even three cars, and one car is downright unthinkable. I don't believe anybody else is going to cut back either. There's a reason, and it's not just that Americans are car crazy (though they are). It's the freedom, convenience, and flexibility that come from having a car at your disposal. The automobile is the most freeing instrument yet invented. It allows folks to take jobs far away from their homes. It enables them to live far from central cities and, if they're anti-social, far from other people. Two cars make the two-earner family possible. And the attachment has grown, as cars have become an extension of home and office, with telephones, message pads, coffee cups, books (on tape), etc.

The European experience is not as extreme as this, nor is it so uniform, as there are more obvious contrasts within the continent. However, the processes involved are the same:

- fundamental changes in the location of economic activity
- increases in wealth and car ownership
- the development of urban highways and commuting transport systems

Cities in NICs

Urban sprawl has also taken place in these cities, but it is much more constrained. The hinterlands of most US cities have extremely low population densities and are farmed at relatively low levels of intensity. By contrast, the rural areas around many Southeast Asian cities are densely populated and intensively farmed. This difference means that land for road construction is relatively more expensive than in the US. In Japan, which is also intensely cultivated and densely populated, $0.70 of every dollar of road investment is spent on land acquisition, compared with an average of less than $0.25 in the US. This difference in the relative value of land (due to the economic intensity of its use) makes the economic cost of a high level of private motor vehicle use much higher in Asia than in the US and in other regions.

The social and environmental costs of this sort of suburbanisation are also much greater in Asia. The land consumed by tarmac in suburban areas around Jakarta (Java) displaces more than 100 times as many people as in the US. If all of the roads which the World Bank claims are justified by economic analysis on the island of Java are actually built, some 800 000 people could be displaced. In Bangkok, roughly 20 000 people a year face eviction from central areas and about half of these evictions are due to road projects. This process of displacement is forcing an increasing percentage of the urban poor to relocate to distant suburban areas. Meanwhile, an estimated 250 km^2 of agricultural land, forest or wetlands are converted every year to urban uses on the island of Java, with enormous environmental consequences.

Cities in the poorer LEDCs

The processes here are quite different. For the most part, the dominant driving force remains rural–urban migration. The essential elements of this are well known. On the margins of many cities in LEDCs, there are recognisable areas of:

- squatter settlements
- spontaneous settlements

The distinction between these terms needs clarifying. Inhabitants of **squatter settlements** have no legal right to the land on which they build. They are not the owners, but the settlements often have an element of planning and infrastructure. **Spontaneous settlement** is a reference to unplanned building, usually without any infrastructure of water, sewerage, electricity or transport being established.

Rio de Janeiro, like many LEDC cities, is experiencing a dramatic increase in population. This increase has come mostly in the form of the rural poor migrating to the cities. Because of the high land values and the enormous demand for space, these poor are forced into squatter settlements known as *favelas*. Named after the location of the first such settlement, a hill called Morro da Favela, these settlements usually occur in two areas of Rio:

- along the steep hillsides otherwise avoided by housing
- along the outer fringes of urban expansion

Greenbelt policies

There are attempts to restrict urban development. This is commonly known as **greenbelt** policy.

In France, urban growth boundaries are used for two main purposes: to rein in urban sprawl by promoting more compact development; and to protect wildernesses, agricultural land and other resource land from low-density, poorly-regulated development. Greenbelt policies in the UK, which attack the same problem less comprehensively, function by setting land aside and by putting some restrictions on development within the belt — hence the leapfrogging of the belts that has taken place in the UK. In the US, the more simplistic idea of a 'belt' has generally been replaced with the notion of corridors of open space and boundaries to development.

The influence of urban economies on the socio-economic characteristics of rural areas

It is obvious that urban influences extend beyond the limits of the urban area. These influences are both social and economic. Within MEDCs, influences might include:

- impact on the social character of rural settlements
- resultant changes in provision of services and commerce
- growth of second-home ownership and seasonality of populations
- increased number of commuters
- changing employment structure within rural areas

The distance from major urban centres and the relative importance of these centres in the urban hierarchy affect the extent of these changes. Thus, the villages close to a major world city such as London or Paris are much more profoundly affected than those that are an equivalent distance from lesser cities.

There are also profound changes in LEDCs. These include:

- rural–urban migration
- the demographic impact on rural areas of this migration
- the impact on agriculture of increasing demand for food from rapidly growing cities
- the growth of an informal manufacturing sector in urban economies, with linkages to rural settlements

Development processes

The world is characterised by wide variations in development

Development can be understood on a number of levels, including social, economic, cultural and political

Development is a complex term. A preliminary explanation of some popular terms is necessary. These include:

- developed and developing countries
- North and South
- First World, Second World and Third World
- core and periphery

Developed and developing countries

The terms **developed** and **developing** have fallen into disuse. 'Developed' suggests that the processes involved have stopped operating; hence, a developed project is something that is completed. It seems odd to apply this to a country or region that is obviously still changing, growing and developing! By extension, 'developing' is applicable to all countries, except those that are becoming poorer or less developed.

The terms 'more economically developed country' (MEDC) and 'less economically developed country' (LEDC) have become common. However, in order to recognise the difference between these, it is important to realise that there is a continuum here from more to less (in the same way that there is from fat to thin). In addition, the terms depend on the context. Thus, a fat friend does not look fat when surrounded by sumo wrestlers; similarly, an MEDC like Portugal may have more similarity to an LEDC like Algeria than it does to the US or Canada. These are, in other words, *relative* concepts.

North and South

The North/South division of the planet is not exact, given the position of the equator, but it signifies an economic division and was popularised by the **Brandt Report**, published in 1980.

The Brandt Report sought a balance in development policies and demanded that the countries of the South be integrated into the global economic system. The authors argued that this would bring about needed improvements in economic and social conditions in disadvantaged countries. At the same time, the rich industrial countries of the North were called upon to share their means and power with the countries of the South. The report contained a number of proposals for the reform and transformation of the world economic system. It concluded that the introduction of such a new system would be an important contribution to the survival of humanity. The connection between armaments and poverty in the countries of the Third World was pointed out.

First World, Second World and Third World

The use of the term Third World has also been common since 1945. Its origins were in a division of the world into political groupings: the First World, which was capitalist, and in which the USA was the dominant military power; the Second World, which was socialist or communist, and in which the USSR was the dominant power; and the Third World, which was made up of non-aligned countries over which the first two worlds competed for influence.

Core and periphery

The use of the terms '**core**' and '**periphery**' is common. It refers to a controlling, powerful and usually rich core and a poorer and frequently **dependent** periphery. The processes that interconnect these two are complex and highly controversial.

A further complication to all of these categorisations is that there is a natural tendency to ignore those variables or factors that cannot easily be measured. We might agree that human happiness or freedom of expression is a fair indicator of development, but neither of these is measured easily. This important qualification needs to be born in mind when tackling the following section.

Investigative work to measure development using a range of indicators, including the Human Development Index

The last section introduced some of the problems of defining development. It is generally accepted that development should include variables that cover the **economic**, **social**, **cultural** and **political** aspects of a society.

Unfortunately, neither the cultural nor political variables are easily measured (quantified) and tend to be ignored, except as footnotes. The usual measures are dominated by social and economic variables. The **Human Development Index** (HDI) is the most commonly used, although there are other initiatives in the area of statistics. New surveys have been devised to measure different aspects of human development, including the Living Standard Measurement Study (LSMS) and the Core Welfare Indicators Questionnaire (CWIQ) surveys, both supported by the World Bank. The Demographic and Health Survey (DHS) is mostly financed by US-AID and the Multiple Indicator Cluster Surveys are sponsored by UNICEF and WHO.

The data used on such measurements are not perfect. Obvious problems include:
- the comparability of economic data from country to country
- the quality and reliability of the initial data
- the difficulty of measuring economic activity that is not official or legal (e.g. the black economy or subsistence production)

- the variations within a country, making some averages very unreliable (e.g. per capita income)

Following the lead of other nations, a few years ago the USA replaced GNP (**gross national product**) with GDP (**gross domestic product**) as a measure of the value of the nation's output of goods and services. The difference between them is that while the latter measures all the output produced domestically (within the borders of the USA), the former does not, as it only measures the output of firms owned by Americans. However, the GNP includes the output of American-owned firms located abroad, while the GDP does not. Due to the fact that, in recent years, domestic savings have been inadequate to finance the desired level of domestic investment, GDP and GNP may in the future increasingly diverge because of a resulting rise in foreign ownership of firms located in the USA. GDP is the best known feature of the national income. It is a summary and quantifiable measure of national well being. Many economists complain that the government spends too little on collecting this kind of data and, as a result, it is not adequate or accurate enough, even in a country as rich as the USA.

The table below displays the usual component parts of the HDI and highlights variations between the different indicators. A correlation test such as Spearman's can test the relationship between these indicators, especially the economic measures and the social indicators. The last column, showing the difference between the key economic indicator of gross domestic product per capita and the HDI rank, is particularly instructive.

Human Development Index rank	Life expect-ancy	Average 1st, 2nd, 3rd grade school enrolment ratio	Real GDP per capita (US$)	Adult literacy (%)	Human Develop-ment Index	Real GDP per capita rank minus HDI rank
1 Canada	79.0	99	22 480	99	0.932	12
2 Norway	78.1	95	24 450	99	0.927	5
3 USA	76.7	94	29 010	99	0.927	0
4 Japan	80.0	85	24 070	99	0.924	5
5 Belgium	77.2	100	22 750	99	0.923	6
6 Sweden	78.5	100	19 970	99	0.923	18
7 Australia	78.2	100	20 210	99	0.922	15
8 Netherlands	77.9	98	21 110	99	0.921	9
9 Iceland	79.2	87	22 497	99	0.919	3
10 UK	77.2	100	20 730	99	0.918	9
11 France	78.1	92	22 030	99	0.918	9
12 Switzerland	78.6	79	20 150	99	0.914	−6

Development can be defined in terms of its sustainability over time

This is another topic with some disagreement over basic concepts. However, few would disagree that **sustainable development** can be defined as:

> ...meeting the needs of the present without compromising the ability of future generations to meet their own needs.
>
> (World Commission on Environment and Development)

Most approaches to this topic contain a series of important ideas with which you will need to be familiar.

- The **carrying capacity** is the maximum number of individuals of a defined species that a given environment can support over the long term at a reasonable quality of life. The idea of limits is fundamental to the concept of carrying capacity. For human beings, calculating carrying capacity is problematic, since we alter our environment and our expectations of a reasonable quality of life. Some argue that the concept is meaningless, as free market conditions and technological innovation can extend limits indefinitely.

- A **steady state economy** is an economy characterised by constant population, capital stocks and rate of material/energy consumption and production such that there is sustainable balance between human activities and the environment. Hence, a distinction is made between growth (which is measurable) and development (which is not always measurable).

- An **ecological footprint** is the area of land and water required to support a defined economy or population at a specified standard of living. Industrialised economies are considered to require far more land than they have and thus, through trade, they impact on resources in other countries. Also known as 'appropriated carrying capacity', this concept includes the distributional aspects of sustainable production and consumption.

- **Natural resources (or capital)** are an extension of the economic notion of capital (manufactured means of production) to environmental goods and services. It refers to a stock (e.g. a forest) which produces a flow of goods (e.g. new trees) and services (e.g. erosion control or habitat). Natural resources can be divided into renewable and non-renewable; the level of flow of non-renewable resources (e.g. fossil fuels) is determined economically and politically.

- **Environmental debt** is the cost of restoring previous environmental damage as well as the cost of recurrent restoration measures. Unless measures are taken to prevent environmental degradation, environmental debt is bound to rise and the debt is transferred to future generations. However, some environmental damage, such as species extinction, is not restorable, and therefore cannot be included in the environmental debt even if it were readily measurable.

- **Inter-generational equity** is the principle of fairness between people alive today and future generations. The implication is that unsustainable production and consumption in the present will degrade and destroy the ecological, social and economic basis for our children in the future; whereas sustainability involves ensuring that future generations will have the means to achieve a quality of life equal to or better than today's.

- **Intra-generational equity** is the principle of equity between different groups of people alive today. It implies that consumption and production in one community should not undermine the ecological, social and economic basis for other communities to maintain or improve their quality of life.

The complex relationship between natural resources and development

This is not a very obvious relationship. At an intuitive level, it seems obvious that the more resources a country has, the better off it will be. However, both on a global and local scale, the relationship is far from clear. Three types of resources are generally recognised:

- natural resources
- material resources
- human resources

Natural resources are those things found in the natural world that are useful to us, and which we have the ability and the desire to use.

Material resources are the stocks of capital held in terms of machinery, equipment and the built environment.

Human resources are the abilities and potential of the human population in terms of our educational levels, skills and capacity to innovate and invent.

It is evident that development is dependent on a combination of these resources. Some countries have experienced very rapid economic development despite a lack of natural resources, for example Japan and the Netherlands. On the other hand, some countries with abundant natural resources have failed to experience equivalent rates of economic development (e.g. Brazil).

We cannot explain the relative economic performances of Japan and Brazil in the past 50 years in terms of their very different levels of natural resource endowment. The most productive way of looking at this is to view natural resources as a variable that is not central to economic growth in the capitalist world. (It was more important in a world without significant trade and exchange.)

Case study: Uruguay

It could be argued that natural resource endowment can obstruct balanced economic development because it leads to too great a dependence on one or more natural resources. This discourages industrialisation and the development of the other two main categories of resources, because the raw materials are sufficient to generate high incomes as long as they last. The history of supply economies in Uruguay illustrates this:

- Growth rate, 0.77%
- Birth rate, 17.42 per 1000
- Death rate, 9.06 per 1000
- Infant mortality, 14.14 per 1000 live births

- Fertility rate, 2.37
- Literacy, 97.3%
- A migrant population arrived largely from Italy and Germany at the end of the nineteenth century.
- Small-scale farmers settled in the interior on a few large estates.
- Labour was in short supply; wages were therefore high and there was a tendency towards family farms.
- Income was distributed evenly, with no great distortions of wealth.
- A pastoral economy developed, specialising in the production of beef and sheep for export.
- Beef, leather and wool were exported, largely to Europe.
- The economy flourished in the early twentieth century and was held up as a model of Latin-American development.
- Wealth was re-invested in the welfare system, health service and infrastructure.
- Uruguay was regarded as the 'Switzerland' of South America (being small and rich) and was rated 11th in the world wealth table in the 1930s.
- Montevideo became the capital city and only urban centre (with a primacy of × 20).
- The city was devoted to administration, trade and central place functions of exchange.
- Problems began after the Second World War.
- A loss of markets followed the formation of the EEC.
- Competition arose from Australia and New Zealand.
- Substitute products (e.g. artificial fibres) emerged, causing markets to disappear.
- Crash industrialisation began in the 1950s, but with no previous industrial tradition. This involved high costs of starting up from scratch.
- Catastrophic failure resulted, which was made worse by growing national debt, leading to cutbacks in welfare and education and increased taxation.
- An outbreak of civil war in the late 1960s (the rise of the Tupamaros) resulted in 1 million people emigrating to Brazil.
- The army took over the country in 1973.
- Uruguay reverted to an export-oriented economy, specialising in cheap, semi-processed products (e.g. leather goods with markets largely in Brazil and Argentina).
- Uruguay slid down the league table of wealth and development.

Within countries there are wide variations in economic growth and development

Regional variations in economic, social and political development

This section is concerned with regional development, uneven development and spatial inequality. The most serious problems facing humankind at the moment are gross variations from one area or region to another, in terms of life opportunities and the impact that economic development is having on the biosphere. This section tackles both of these problems.

In **classical economic theory**, regional differences should not persist because regions with a high demand for labour will ultimately become high-cost regions, whereas regions that may start poor will become cheaper areas for industry as demand falls. The hidden hand of the market was supposed to even out differences. For some, this was reason enough to disregard regional policy. However, regional differences have persisted in many countries and it was Gunnar Myrdal who provided the first coherent explanation of this.

Myrdal's cumulative causation model

Myrdal suggested that regional differences are due to the following:

- Suppose there are two regions with one (Region A) being more developed and populated than the other (Region B), perhaps because of natural resource advantages or a preferable location.
- Firms in Region A will have larger markets and be able to realise **economies of scale** not available to firms in Region B. Thus, firms in Region A may have a cumulative and growing advantage over firms in Region B. With higher incomes in Region A, there is an incentive in that region for a shift into more advanced production, further increasing the income gap. This may also induce technical change, generating further divergence between the two regions.
- Region B may be limited to producing goods for which the **income elasticity of demand** is low (as income rises, the demand rises more slowly, as with agricultural goods and primary products). Demand for the Region B products may be reduced by the development of substitutes in Region A. Better investment opportunities in Region A may attract the more efficient elements of capital and labour away from Region B, further limiting its production possibilities. These effects are called **backwash effects**.
- On the other hand, Region A will also have a growing demand for the resources of Region B and the ability and wealth to exploit and control these resources. This is a stimulus for the development of Region B, but the profits of resource extraction may largely go to Region A. A local elite may arise who cooperate and help control this process in exchange for a share of the benefits.
- Other **spread** or **trickle-down** effects may include an improved transport system, so that the resource(s) can be exported at lower costs. This may also allow Region B to export its own products at lower costs. Competition from imports from the more efficient Region A may lead to the closing down of the least efficient enterprises in Region B. This will cause productivity in Region B to rise and allow the employed to earn higher incomes. In the longer term, increasing congestion and pollution in Region A, coupled with rising factor costs, may cause some economic activities to relocate from Region A to Region B.

Today, the nature of the world economy is such that local problems must be seen in the context of a global economy.

Growth poles

Growth poles are locations developed explicitly with the intention of stimulating remote regions in an attempt to even out regional disparities and avoid some of the imbalance outlined by Myrdal. They have usually involved major investment from government, which often develop the infrastructure, with private capital being encouraged to join in. The core of these poles has usually been large-scale industrial development, such as a hydroelectric project or the exploitation of a previously neglected natural resource supply. The theory is that once established, these investments lead to the **multiplier effect**, attracting other industries until **sustainable development** is achieved. This often involves a major regional development plan.

Reasons for, and the impact of, regional development policies

Regional development policies aim to improve the economic performance and living standards of specific regions or areas of a country. They operate on a number of scales from local (e.g. Cardiff Bay and the London Docklands) to regional (Highlands of Scotland and Carajas in Brazil). There are two main types:

- The '**top–down**' approach, led by national governments or trading unions (e.g. the European Union), usually involves large-scale projects with major investments in infrastructure and a concentration of resources on a **growth pole** or **growth corridor**.
- The '**bottom–up**' approach usually takes a small-scale **people-centred** approach, draws funds from charities and is often highly local (e.g. the well project in The Gambia). These projects aim to be **self-sufficient** but still need to be integrated into wider (and more expensive) national policies.

The theoretical background is provided by Myrdal's ideas on cumulative causation on the one hand and classical theory (as last practised in the UK by the Conservative administration of the 1980s) on the other. Key ideas include the concept of the **multiplier effect** (both positive and negative) and **spread effects** (or **trickle-down**). In general, the principle of **interventionism** has been accepted by almost all national governments since the 1930s.

At least three case studies would be useful. It would be sensible to draw one from an LEDC (e.g. Carajas — see below), one medium/local scale from an MEDC (e.g. south Wales and Cardiff Bay) and perhaps a small-scale **intermediate technology** (bottom–up) example from an LEDC (e.g. The Gambia).

There is considerable debate about the effectiveness of these schemes. You should run through the costs and benefits involved and the problems that they have faced. Look particularly at the **externalities** of these schemes. You should not be afraid to suggest that many of them have only worked for a limited group of investors and TNCs without bringing much long-term benefit to the region in terms of balanced and sustainable growth. However, do not underestimate the real benefits that have stemmed from some of these schemes. What follows is a guide to the level of detail that would be useful for an essay on this topic.

Case study: the Grande Carajas Programme

- The Grande Carajas Programme is one of the largest integrated development schemes ever undertaken in tropical rainforests.
- It covers an area of 880 000 km², almost 11% of the country of Brazil.

- Major schemes include iron-ore mines, two aluminium plants, a hydroelectric power plant and the extension of roads and railways.
- One of the most controversial aspects of this project is its scale of environmental degradation in the Amazon. The deforestation of Amazonia leapt from 2.4% in 1978 to 10% in 1988. At the heart of the area, the figure is estimated at over 30%.
- Another important criticism is about social impact on rural peasants and indigenous people. It has involved huge population displacement and has accelerated the disabilities of food supply.
- On the other hand, the project generates and processes 45 million tonnes of iron ore annually, as well as 1.5 million tonnes of manganese and 10 000 kilograms of gold. It is also preparing to produce copper, bauxite and other metal ores.
- All these activities are constantly monitored by an environmental policy that, officially at least, strives to conserve and regenerate the delicate Amazon ecosystems.
- The plan was finally approved in 1980, and the World Bank, Japan and EEC agreed to lend money. Japanese advisers played a particularly significant role in persuading the Brazilian government, and produced several influential reports. They started aid even though none of the twelve smelters approved by 1987 had carried out the environmental impact assessment required by the World Bank.

Advantages and challenges of core and peripheral regions, to include positive and negative multiplier effects, and spread and backwash effects

The key concepts in this section have already been covered in previous sections. You should be able to apply these ideas to two contrasting regions from the same country. A review of regional policy in the **peripheral** region of southern Italy, the **Mezzogiorno**, is given below.

Case study: the Mezzogiorno region of Italy

After many years of effort, it became clear that attempts to improve agriculture had not reduced unemployment, raised incomes or narrowed the overall economic gap between the south and the rest of Italy.

Many experts had insisted all along that only rapid industrial development would create the jobs and prosperity needed to break the vicious circle of unemployment and poverty. Put simply, industry was to be moved to the area of surplus labour, rather than labour moving to an area of industrial activity. It was hoped that this would slow down south–north migration.

Since there was no southern money for such industrialisation, state investment funding shifted from agriculture to industry. Special **growth poles** were designated in the Naples–Salerno area; in the Taranto–Bari–Brindisi region, a triangle of giant basic industries was to be established. It was believed that these would in time attract further, secondary industries and create a broad and diversified industrialisation through the operation of the **multiplier effect**. The south would then be on its way to autonomous and **sustainable** economic development.

The development fund's spending on **infrastructure** was cut from 42% in 1957 to 13% in 1965, while assistance to industrial enterprises was stepped up from 48% to 80%.

To encourage private investment, the government offered generous grants and tax incentives. It also required the state holding companies, IRI and ENI, to place increasing amounts of capital in the south. IRI, the major investor, injected over $20 billion into the Mezzogiorno between 1963 and 1978.

In this way, a steel industry (at one time the most modern in Europe) was developed at Taranto as well as an Alfa Romeo plant near Naples and petrochemical factories in Sicily and Sardinia. The volume of industrial development in the south during this period increased sevenfold.

Even though some of these investments were productive and profitable, the large-scale heavy industrialisation programme did not turn out to be the solution to the south's problems, any more than it was in Uruguay and in many other cases. The new industries were not labour-intensive and created relatively few jobs.

The managerial and technical personnel were recruited mostly from the north and eventually returned there, leaving the south without a local managerial and technical class, much as Myrdal had argued would happen.

The industries themselves were subsidiaries or **branch plants** of northern companies, which never developed links to the local economy.

Laws were therefore enacted in 1971 and 1976, requiring the development fund to concentrate on a new range of projects. These included the further development of industrial infrastructure, the exploitation of natural resources, social projects in metropolitan areas, water-reclamation projects, cleaning the Gulf of Naples, extending the road network into remote mountainous areas, reforestation and promotion of citrus fruit production.

Many of these programmes had to be scaled down or scrapped as a result of the economic crisis following the oil price increase after 1973. Some of them were gigantic mistakes. Over $800 million was spent on cleaning the Gulf of Naples without producing a litre of unpolluted water.

By the end of the decade, the era of big infrastructure projects was over. Emphasis then shifted to the selective encouragement of small- and medium-sized firms, which in the early 1980s accounted for over 90% of the projects and 85% of the investment of the development fund.

The fund, originally established for 15 years, was extended year after year until 1984, when it was finally terminated. During the first 30 years, between 1950 and 1980, it spent about $20 billion of the total $30 billion allocated to it. The European Community's regional fund added over $2 billion of its own, making the Mezzogiorno its single main beneficiary.

Many Italians remain doubtful about the benefits of these policies. The south remains a poor region relative to the rest of Italy, and the difference between north and south remains large. There are clear advantages in some areas, but they are not spread evenly in the region.

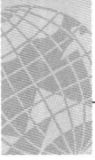

Development is a process that changes through time

The economic, social and cultural factors affecting the rate and nature of development

This is a substantial and highly controversial topic. You need to have a grasp of the theory and an understanding of the terminology, as outlined below. You also need at least two case studies with which to illustrate your understanding of development in practice.

Economic development is the analysis of the economic progress of nations. It encompasses a mixture of geography, sociology, anthropology, history, politics and, importantly, beliefs. It used to be known as political economy.

Studies of **economic development** began in the 1930s when it became evident from the colonies that many people did not live in an advanced capitalist economic system.

Early **economic development theory** was an extension of conventional economic theory that equated **development** with growth and industrialisation. As a result, Latin American, Asian and African countries were seen mainly as **underdeveloped** countries, i.e. primitive or backward versions of European nations that could, with time and help, develop the institutions and standards of living of Europe and North America. The analogy with human development from embryo to adult was often made.

As a result, **Rostow's stage theory model** dominated discussions of economic development. This crude model argued that all countries passed through the same historical stages of economic development and that underdeveloped countries were merely at an earlier stage in this linear historical progress, while First World (European and North American) nations were at a later stage. Rostow pointed out that along this development path there was a risk of falling into the trap of communism, which would prevent full development.

More enlightened attempts led to the conclusion that while there were no obvious linear stages, countries tended nonetheless to exhibit similar patterns of development, though differences could persist. The job of the economist was to suggest short cuts by which underdeveloped countries might catch up with developed nations.

Capital formation through savings was recognised as the key to accelerating development; by increasing individual savings, money would be available through the banking system for industry to buy machinery and thus escape the **vicious circle of poverty**. The poor do not save and do not consume much, so there is no demand for manufactured goods. Thus, there is no wage economy and therefore people remain poor. Many argued that countries required outside assistance to escape this vicious circle.

The early work on the **dual economy** by Lewis emphasised the role of **savings** in development. Many modern observers have noted the very high rate of saving in Asian economies such as Taiwan and Korea.

Savings were encouraged by governments, making government-directed industrialisation feasible, and laying the foundations for **growth pole theory** and Myrdal's theory of **cumulative causation**. Thus, **government involvement** (whether by planning, socio-economic engineering or effective demand management) was regarded as a critical tool of economic development.

Other economists saw international trade as the major stimulant to growth, arguing that increasing specialisation and free trade would bring benefits to all. The **law of comparative advantage** underpinned this view.

Although capital formation and investment remained the central part of these theories, an increasing amount of attention was paid to human resources, leading to an emphasis on **education** and **training** as prerequisites of growth and the identification of the problem of the brain drain from the LEDCs to MEDCs.

Social development as a whole, notably education, health and fertility, would, by improving human capital, establish the foundations for growth. According to this view, industrialisation, if it came at the cost of social development, could never be **sustainable**.

By the 1960s, development was taken to include the elimination of poverty, unemployment and inequality, as well as economic growth. Thus, structural issues such as dualism, population growth, inequality, urbanisation, agricultural changes, education, health and unemployment began to be treated as a central part of development.

An important debate on the very desirability of growth was stimulated by **Schumacher** in his book, *Small is Beautiful* (1973). He argued against the desirability of industrialisation and extolled the merits of handicraft economies. As the world environmental crisis became clearer in the 1980s, this debate took a new twist as the very **sustainability** of economic development was questioned. It became clear that the desirability of development needed to be reconsidered.

Another idea that became widely known at the same time was the **structuralist** thesis. This called attention to the distinct structural problems of LEDCs and argued that they were not merely primitive versions of developed countries — they had distinctive characteristics of their own.

One of these characteristics was that, unlike MEDC industrialisation, LEDC industrialisation was supposed to occur while these countries existed alongside already industrialised MEDCs and were tied to them by trade. It was argued that this could give rise to distinct structural problems for development.

Dependency theory in economic development suggested that the world had developed into a **core–periphery** relationship among nations, with LEDCs being forced into becoming the producers of raw materials for MEDCs and thus condemned to a peripheral and dependent role in the world economy. The logic of this theory was that some degree of protectionism in trade was necessary if these countries were to enter a self-sustaining development path. **Import substitution**, enabled by protection and government policy, rather than trade and export-valorisation, was the preferred strategy. Historical examples of government-led industrialisation, such as Japan and Soviet Russia, were held up as proof that there was not simply one path to development, as had been implied by the stage theories of Rostow.

Many LEDC governments adopted the language and policies of the structuralists and/or the **Marxists** who supported this broad structuralist view in the 1960s and 1970s. **Neo-colonialism**, **core-periphery** and **dependency** were the keywords of this period.

The **neo-classical** thesis was simple: government intervention did not only impede development but actually thwarted it. The emergence of huge bureaucracies and state regulations, proponents argued, suffocated private investment and distorted prices, making developing economies extraordinarily inefficient. In their view, the ills of unbalanced growth and dependency were all ascribed to too much government interference, not too little.

In recent years, the neo-classical thesis has gained greater support, particularly in Latin America. However, the evidence is still ambivalent and disputed. Both structuralists and neo-classicists use rapid East Asian development and the subsequent crisis, as well as the disastrous African experience, as proof of their directly opposing theses.

The rise of the newly industrialised countries (NICs) in East Asia

Forty years ago, anybody predicting the social and economic progress that has been achieved in East Asia would have been regarded as insane. Average per capita incomes in South Korea, for example, were lower than in Zaire or Afghanistan. The level of poverty in Indonesia and Malaysia was similar to that in much of sub-Saharan Africa and South Asia, with both countries dependent on exports of primary commodities such as rubber and copper. Indonesia was a particularly bad case, with huge debts, receiving massive aid and dependent on imported food. The parallels with sub-Saharan Africa today are difficult to avoid. Many commentators saw these countries as beyond hope of foreseeable development and a major threat to world stability. The region became more and more backward.

What has happened since is an astonishing story of economic and social development. The **Asian tiger** economies of Hong Kong, Singapore, South Korea and Taiwan, followed by many other **tiger cubs** in the region, including Malaysia, Indonesia and Thailand, have grown at unprecedented rates, sustaining growth in GNP at 10% per annum over many years. This represents a major achievement and a challenge to those who argued that development was unlikely (if not impossible) in this region.

The **World Bank** and the **IMF** claim the region's record over the past 40 years as proof of the success of free-market policies of the type associated with its **structural adjustment programmes (SAPs)**.

In fact, the opposite is true. A key feature of economic policy in most of the East Asian countries is a rejection of the simple free-market models of the neo-classicists. Many of the policies associated with structural adjustment are inconsistent with the policies that have achieved rapid growth and poverty reduction in East Asia.

The reality is that no single East Asian model exists. The countries of the region have followed a wide variety of policies, reflecting their individual historical, political and economic circumstances. With varying degrees of success, most have combined growth with equity and poverty reduction. But different countries have followed different routes — and they offer different lessons.

Taiwan did not follow South Korea, Indonesia did not follow Taiwan, and China and Vietnam have not followed Malaysia. Each country has developed its own policies and plans depending on its own circumstances. There are lessons from each country, but they vary through the region, and they vary over time.

Three major generalisations emerge from the range of national experience in East Asia:
- Poverty is not inevitable.
- Poverty is best tackled with economic policies that aim to benefit all members of society.
- Governments have a major role to play in social and economic development.

Positive and negative consequences of development
The **positive** consequences of development are that people become better off, fewer people live in poverty and all the measurable indicators of human progress improve. The way in which these benefits are distributed within a society is important. An elite group might benefit hugely from some types of development, but at the expense of the continuing poverty of the majority of the population.

Neo-colonialism
At the end of the nineteenth century, much of the world was divided into empires which consisted of the home or mother country and a number of colonies. The function of these colonies was to produce raw materials for the industries of the mother country.

This colonial system provided the structure for the modern world system. The colonies exported one or two **primary commodities** (raw materials) which were very **low value-added** goods, such as tea (from Kenya) or copper (from Zambia), and imported **higher value-added** manufactured goods, such as electrical wire.

Some economic historians have argued that the mother country often prevented industrialisation from taking place in colonies, or broke up existing industry if it posed a threat to the colonial power. The Indian textile industry is frequently quoted as an example of the latter.

Whatever the shortcomings of colonial rule, the overall effect was positive in reducing the economic gap between Europe, the USA and the rest of the world. It has been argued that this brought enlightenment where there was ignorance. It suppressed slavery and other barbaric practices such as pagan worship and cannibalism. Formal education and modern medicine were brought to people who were seen as having limited understanding or control of their physical environment. The introduction of modern communications, exportable agricultural crops and some new industries provided a foundation for economic development. Africans, for example, received new and more efficient forms of political and economic organisation.

In the post-war period (from 1945 to the present) almost all of the former colonies were given their independence — at least in the sense of being able to govern themselves.

Critics argue that **neo-colonialism** involves the adoption of new methods of extracting wealth from the former colonies by using **TNCs** as agents and, most commonly, **foreign direct investment** as the weapon.

Today, for example, Africa owes $227 billion to Western creditors. This crushing debt burden is keeping the continent impoverished. These countries are tied into a system that more or less compels them to take on a **dependent** role in the world economy, concentrating on the production of raw materials and cheap commodities, much as they did in colonial times.

Dependency

Dependency theory arose in the 1960s and argues that the growth and development of the MEDCs is, ultimately, dependent on pumping out the wealth of the LEDCs. In other words, the MEDCs develop through **underdeveloping** the LEDCs. Under-development becomes a process rather than a description of an economy. These LEDCs are dependent upon the MEDCs. The MEDCs are dependent upon the process of transferring wealth from the LEDCs. Despite the rise of the NICs, which challenges this view, the evidence for the supporters of dependency theory is quite strong.

The gap between the MEDCs and LEDCs is more pronounced than it has ever been. The United Nations *Human Development Report* for 1997 shows that the share of world trade for the 48 least developed nations, representing 10% of the world's population, halved in the previous two decades to just 0.3%, with over 50% of all developing countries not receiving any foreign direct investment (two-thirds of which went to just eight developing countries).

In fact, around 100 LEDCs are experiencing slow economic growth, stagnation or out-right decline, and the incomes of more than a billion people now no longer reach levels attained 10 or even 30 years ago.

The same report indicates that 1.3 billion people live on a dollar a day, or less; that there are 160 million malnourished children; that one-fifth of the world's population is not expected to live beyond 40 (in some countries, life expectancy has fallen by five years or more); and that 100 million people in the North live below the poverty line (the North also has 37 million jobless people). Over a billion human beings lack access to safe water, nearly a billion are illiterate, and around 840 million experience hunger or food insecurity.

The report also shows that the net wealth of 10 billionaires is worth 1.5 times the combined national incomes of the 48 least developed nations.

The disparities between North and South are increasing fast. The share in global income of the poorest 20% of the world's people has fallen from 2.3% in 1960 to 1.4% in 1991 and to a current level of 1.1%, while the ratio of the income of the top 20% to that of the poorest 20% rose from 30:1 in 1960 to 61:1 in 1991, and grew still further to a figure of 78:1 in 1994.

Thus, the view that the current state of the world can be summarised as 'the rich are getting richer and the poor are getting poorer' seems hard to dispute. What dependency theory argues is that these two ideas are linked, i.e. 'the rich are getting richer *because* the poor are getting poorer'. As you might imagine, this is a controversial viewpoint.

The debt crisis

The end of the last **Kondratieff wave** of economic growth is generally dated to the end of the 1960s. At this point, the world economy was facing a difficult period, with too much capacity (too many factories) and too little demand for goods (increasing ownership of cars and household electrical equipment had fuelled a long period of growth, but most who could afford these things had them, so demand had declined).

The price of oil was raised on world markets by 500% after the Arab–Israeli war of 1973. Governments in the oil-producing countries accumulated huge quantities of so-called petro-dollars (profits from oil sales). Since they could not spend them fast enough, they put much of the money into Western banks.

Banks have to lend money to make money; that is their function and their reason for existence. If they do not, they cannot survive. Unfortunately, the depression in the MEDCs meant that neither individuals nor industries were much inclined to borrow money. There was an economic recession, people feared unemployment and industries were experiencing a reduction in profits as demand for their products fell. This negative multiplier effect led to varying, but increasingly severe, levels of economic recession in most MEDCs.

The banks hunted around for 'new' customers and found them in LEDCs where they offered loans at relatively low rates of interest to recycle their large reserves of petro-dollars. Much of this money was poured into large-scale economic development schemes, including major dams, transport systems and crash industrialisation programmes, with the aim of producing enough goods for sale to allow them to pay back their loans from the economic growth that would come from these schemes. Other loans were used to buy armaments or wasted on hugely extravagant prestige projects (international airports were a favourite). Corruption was common and the management of the loans was very poor.

At the end of the 1970s, changes of government in the US and UK ushered in a period of monetary policy that used interest rates as a weapon against price rises. Interest rates rose and in the effort to defeat inflation, demand was cut back further and the recession became a depression. For many LEDCs which had borrowed money, this was a catastrophic combination of circumstances. They were forced to pay higher interest rates on loans that they had little chance of repaying because the markets for their goods were in economic recession and demand was stagnant or even falling, forcing down prices, especially of raw materials.

The most spectacular and alarming scene of this nightmare scenario took place in the summer of 1982 when Mexico, which owed $80 billion but had no cash to pay the interest let alone repay any part of the loan itself, threatened to default. Nine large US banks had no less than 44% of their entire capital tied up in Mexican loans. If Mexico had defaulted (refused to pay the interest), these banks would have become bankrupt, with untold consequences for the world economy. The crisis was averted with a last-minute rescue package of new loans of $8 billion from the **International Monetary Fund (IMF)**. The first **Structural Adjustment Programme (SAP)** was born.

The debt crisis has not gone away. The recent East Asian crisis, following hard on the heels of the Russian crisis, suggests that recurrent crises might become an increasingly troublesome feature of the world economy. Certainly, the management of that debt has profound implications for all of us.

The role of international links — aid, trade and lending institutions — in the development process

The idea that countries can develop economically and socially without outside links is hard to sustain. **Autarchic development** (as this is known) clearly once took place in the UK (England, to be more precise), as the country was the first to experience an industrial revolution. Even here, however, the linkages with a growing overseas Empire were highly important. The USA can also be cited, given its huge land area and high level of natural resources, rendering it largely independent of the rest of the world should it wish to pursue this path.

For the rest, some degree of linkage is inevitable. These links involve:
- trade in goods
- international aid
- international loans
- flows of capital (**foreign direct investment**)

In the last 20 years, a new global economic order has emerged in which all four of the above play a central role. The key players are:
- the **International Monetary Fund (IMF)**, which is a lending institution. This was set up in 1944 to help stabilise exchange rates and give credit to countries with a balance of payments problem. Its role today is to promote trade liberalisation.
- the **World Bank**, which was set up at the same time to provide loans for the reconstruction of the war-damaged European economies labelled, at the time, developing economies.
- the **World Trade Organisation (WTO)**, which replaced GATT. It implements the agreements made in 1994 by 117 governments in the Uruguay round of negotiations. The thrust of this agreement is a legally binding commitment to free trade.

The funding for the IMF and the World Bank comes from MEDCs. The policies are formulated in similar fashion, with voting rights given in proportion to levels of donation. Hence Europe, Japan and the United States dominate both institutions. The WTO settles disputes through panels of experts and lawyers who make decisions that can be overturned by unanimous vote.

The main policy instrument in the governance of international links is the **Structural Adjustment Programme (SAP)**. This is a set of policies imposed upon LEDCs as a condition of the loans offered to help them repay their debt. The key features include:

- privatisation of state enterprises to try to end inefficiency and to attract foreign investment
- opening of the economy to foreign direct investment and imports by removing any regulations and subsidies, which protect local industries and agriculture from foreign competition
- raising hard currency to help repay loans by promoting exports and tourism
- a reduction of government spending by cutting services, introducing charges for health and education, and cutting back on government employees
- devaluation of the local currency to make exports more attractive by reducing their price on the world market, and increasing the price of imported goods
- increasing interest rates to encourage savings and reduce local spending

Debtor LEDCs have no choice but to accept these conditions, because loans will only be granted to countries that have the official IMF/WB seal of approval. The impact within many of these countries has been to squeeze the poor, especially in countries such as Mozambique and Tanzania, where interest payments exceed the combined health and education budgets.

The search for foreign currency has also involved some sacrifices of the environment, as cash crops are developed for export, while deforestation and open-cast mining result from TNCs moving in to develop new raw material sources.

The terms of trade have continued to move against primary producers, as demand declines when substitutes are found (e.g. fibre-optics for copper wire), or new competitors and over-production force down prices. For example, in 1975, the sale of 8 tonnes of Kenyan coffee could buy a tractor; it now takes 40 tonnes to buy an equivalent tractor. This arithmetic forces many of the poorest LEDCs into a spiral of environmental exploitation that is almost certainly not sustainable.

The role of aid

The boundary between aid and loans that require repayment is somewhat blurred. A common definition of foreign aid is:

> …assistance in the forms of grants and loans given by one government or multilateral organisation to an LEDC, which in the case of a loan must have at least 25% in the form of a grant (or gift). This also includes the costs of providing, for example, expert training or providing advice on economic reforms.

In other words, it is a flow of capital (money) at a concessional rate. The degree of concession will vary greatly between **emergency relief aid**, when both private individuals through official charities (like Oxfam or Christian Aid) and governments give food and medicine or the money to buy it for no charge, to cases when the 20% grant element is almost invisible behind a screen of conditions and restrictions. Aid can also be classified by recipient and donor. **Bilateral aid** is government-to-government assistance, whereas **multilateral aid** has many donors and many recipients.

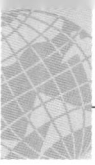

The impact of aid is no clearer than the impact of any other inward investment in LEDCs, whether it be a normal loan or a gift. Some countries, such as Chile, have managed respectable rates of economic growth in recent years despite very low levels of aid. Bilateral aid seldom comes without some political motive. The USA has, by its own statistics, spent over $1 trillion on foreign aid since 1945. It freely admits that the motives for this aid are:

- protecting its political and strategic interests
- promoting US exports
- rebuilding war-damaged economies
- providing relief during humanitarian crises

Only the last of these is entirely altruistic, the other three bringing possible benefits to the donor. Critics of this type of analysis would argue that aid has been repaid many times over. In 1984–85, up to one million people died in the famine in Ethiopia. 'Never again would such a situation be allowed', said many who had campaigned internationally for food aid at that time. Today, Ethiopia has a foreign debt of more than $10 billion, many times greater than the value of its exports. In the mid-1990s, the IMF insisted that Ethiopia implement neo-liberal policies of cutting social spending to ensure budget surpluses. But loan agreements imposed as part of SAPs have seen the interest paid far exceed the amount borrowed, and food production is increasingly for export rather than domestic consumption.

Questions
&
Answers

This section of the guide contains six typical questions for Unit 5, two on each of the three topic areas outlined in the Content Guidance section. In the unit assessment itself, you will have to answer two questions in one and a half hours and you must not answer both questions from the same section. One example of an answer is provided after each question.

Examiner's comments

The examiner's comments are preceded by the icon e. A closing summary is also provided, which indicates the mark the answer would have been awarded. The comments are designed to highlight the strengths and weaknesses of the candidate's answer. The examples provided contain illustrations of typical errors and shortcomings (such as irrelevancy, lack of clarity, lack of focus on the wording of the question or shortage of case study material) as well as evidence of good practice.

Marks

In both parts of the question, marks are awarded according to the level of performance. Each answer is assessed against general and specific criteria and then awarded marks at the appropriate level. An indication of the general criteria is provided for your guidance below:

Part (a)

Lower level	1–3 marks	A basic grasp/awareness evident.
Higher level	4–5 marks	Secure understanding with appropriate exemplification.

Part (b)

Level 1	1–3 marks	Little accurate knowledge. Little relevance to the question.
Level 2	4–7 marks	Mainly descriptive. Poorly written. Simplistic cause and effect. Lack of fluency.
Level 3	8–11 marks	Sound description. Limited understanding. Some use of examples or data. Linked sentences.
Level 4	12–15 marks	Accurate knowledge. Sound understanding. Detailed and varied exemplification. Well-organised answer.
Level 5	16–20 marks	Wide-ranging knowledge. Thorough explanation. Detailed exemplification. Concepts applied. Fluently written.

Economic systems (1)

Study the map below, which shows the Tsukuba Science City in Japan.

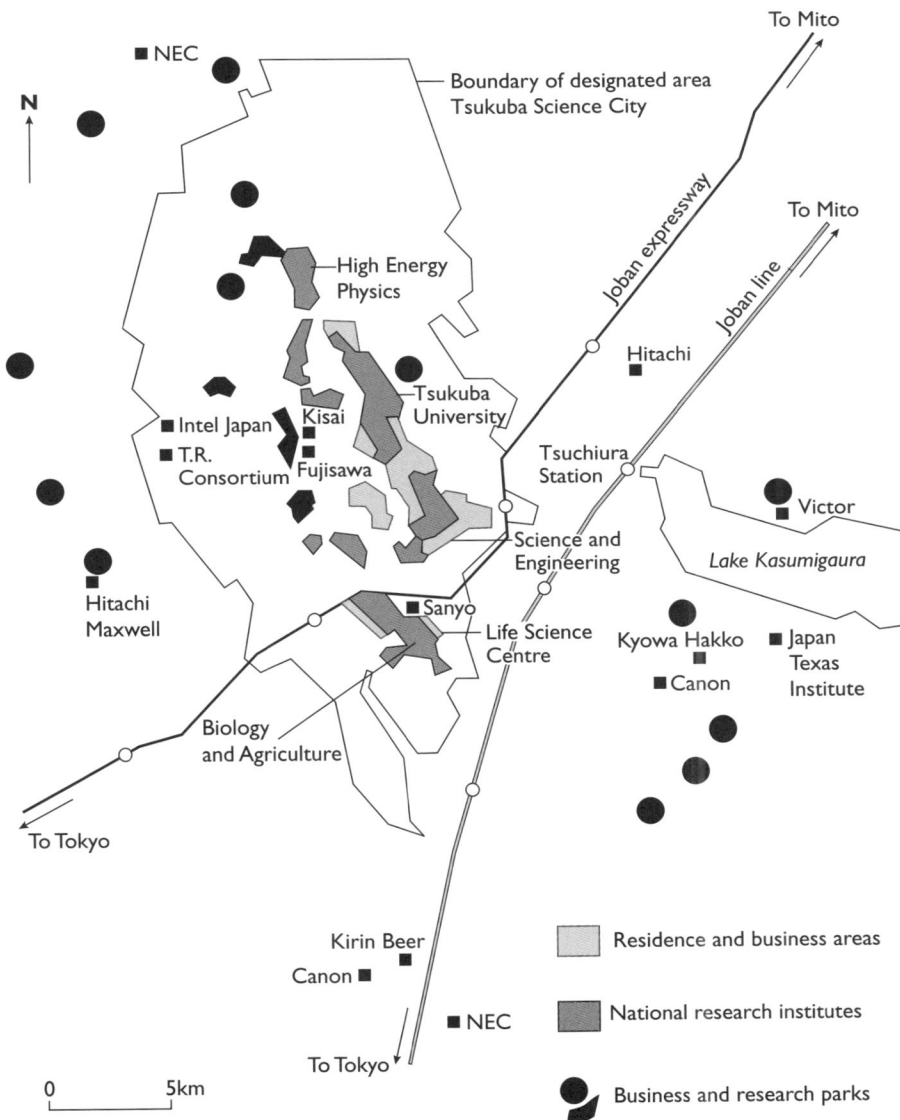

- To Mito
- NEC
- N
- Boundary of designated area Tsukuba Science City
- Joban expressway
- To Mito
- Joban line
- High Energy Physics
- Hitachi
- Tsukuba University
- Intel Japan — Kisai
- T.R. Consortium — Fujisawa
- Tsuchiura Station
- Victor
- Lake Kasumigaura
- Science and Engineering
- Hitachi Maxwell
- Sanyo
- Life Science Centre
- Kyowa Hakko
- Japan Texas Institute
- Canon
- Biology and Agriculture
- To Tokyo
- Kirin Beer
- Canon
- NEC
- To Tokyo
- 0   5km

Legend:
- Residence and business areas
- National research institutes
- Business and research parks

(a) Distinguish between tertiary and quaternary employment. (5 marks)

(b) Examine the causes and consequences of the rise of the quaternary sector. (20 marks)

Candidate's answer to question 1

(a) Quaternary is a new sector of the economy, which is mostly found in MEDCs. It consists of many factors like research and development and has grown rapidly in MEDCs with many consequences, both good and bad. The tertiary sector is the service sector which provides a range of different jobs from bankers through to hospital workers. Like the quaternary sector, nothing is actually made, but generally these are not quite such high-level jobs as those in the quaternary sector.

> 🖉 This is a rather confused attempt to distinguish categories and implies that research and development dominates. The definition of tertiary activity is a little stronger, but no attempt is made to demarcate the two categories. This is a lower-level response, worth 2 marks.

(b) The quaternary sector has arisen due to the research and development of new technology. Examples of this might be computers and mobile phones. But it is not just the development of this type of goods which has created a rise in the quaternary sector; it is also a demand for these goods. Suppliers have responded to a high demand growth in the sector.

> 🖉 This suggests that quaternary activity has arisen as a function of demand for high-tech products and because of the developed technology. It reinforces the confusion between high-tech and quaternary.

The quaternary sector has grown as the rest of the economy has declined in importance, especially in MEDCs like Britain. Jobs in the secondary and primary sectors have disappeared as industries have closed due to foreign competition and changing demand. For example, many jobs in mining have disappeared as we now import much of our coal from Australia. Many of the old uses of coal have also gone, like railways and ships. As a result, new jobs have had to develop. Quaternary jobs are particularly important because they offer the chance of developing new products and goods that will provide employment in the future. The government has helped set up science parks like that at Cambridge where people from the university can work with others to research new types of computers, drugs and bio-engineering. Companies have chosen this location not just because of the university but also because it is quite close to London with all of its amenities and it is an attractive part of England. The same has happened in America where Microsoft has chosen Seattle because of its favourable climate and the attraction of the scenery, including skiing.

> 🖉 This attempts to tackle causes but makes the common error of assuming that as one industry declines, another will arise to take its place. Governments are placed at the centre of the process and some irrelevant locational detail is added.

Growth in this sector of the economy has had many advantages. Demand increases economic activity and growth. This raises the standard of living for people in the country. The people required to work have to be skilled and therefore gain high wages, as there are not many who have the skills. This can be seen in the south of Britain where wages are high and jobs are plentiful, as many quaternary industries are located here. The people who receive these wages will gain an increase in their standard of living. Wages have also become higher in other industries due to demand, so they too see a rise in standard of living. This economic activity means that more people are paying higher taxes, which gives local authorities more funds for education and health. This has been widely seen in southern Britain where education and health have been improved. In areas such as Swindon, new hospitals have been built. These areas will also experience a positive multiplier effect, attracting many new businesses and large amounts of investment. Swindon has seen its old run-down steam railway works turned into a shopping complex.

e This section on consequences is stronger, despite the fact that quaternary industry remains undefined. The use of real places is appropriate and the treatment of the multiplier effect is competent and relevant, though not particular to quaternary growth.

Although there are disadvantages arising from the growth of quaternary industry in MEDCs, industry will locate in an area which is beneficial to it. Therefore, all the advantages tend to be only in one area. This is typical in Britain, with the North falling behind and creating a divide. The rise also leads to a fall in other areas such as secondary manufacturing. This is often seen in the North, with the closure of traditional industries, such as steel in Sheffield, which causes unemployment in the area. Therefore the unemployed take from the local authorities but do not give back. This means fewer funds are available for them and less can be spent on health, education and other projects in the area. With the closure of industry, there will be a downward multiplier effect resulting in a fall in standard of living for the inhabitants. Trying to find a new job will be hard as it is difficult to re-train and even harder to relocate. The downward multiplier means a fall in house prices. It would be harder to sell up and move to the South because of higher house prices. This is one of the great divides of Britain, as a house with four bedrooms and large garden in the North would be equivalent to a one-bedroom flat in London. The unemployed would also have to re-train for the skilled jobs in the South, which takes time. All of these disadvantages require the government to intervene and assist particular areas, a process which may be funded by the higher taxes in the south of Britain.

e This is much less useful than the previous paragraph, although it does demonstrate a recognition that consequences may be negative as well as positive and not evenly spread within a country.

It seems that the quaternary sector arose from the development of new goods and a demand for these goods. Overall, there are many advantages and disadvantages to its growth. Southern Britain appears to have many of the benefits, whereas the North feels the disadvantages.

e The conclusion adds nothing and lapses into truisms.

question

e In general, the candidate recognises the need to tackle both causes and consequences in part (b) and makes some attempt to cover both negative and positive aspects. There is some relevant locational detail, but the lack of clear identification of quaternary industries limits the response to Level 3. It is awarded 8 marks.

Total mark: 2 + 8 = 10/25

Economic systems (II)

Study the map below, which shows the destination of goods produced in Mexico by United
States transnational corporations.

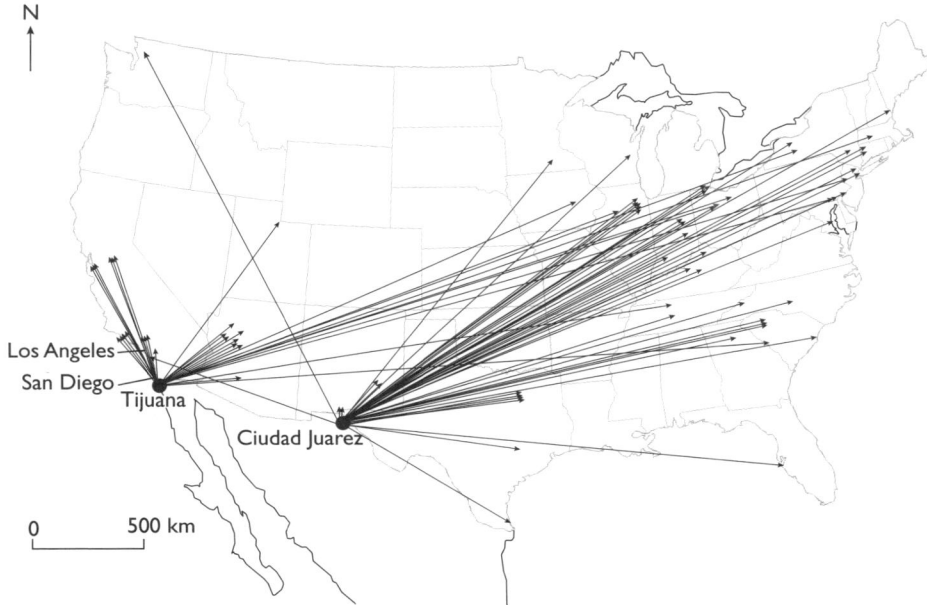

(a) Define the term 'manufacturing industry'.   (5 marks)

(b) Describe and explain the causes and consequences of the changing location
of manufacturing industry.   (20 marks)

■ ■ ■

Candidate's answer to question 2

(a) Manufacturing industries can be divided into two groups: capital and consumer
industries. Capital industries are those which supply material to other manufacturing industries. These industries are relatively insignificant in today's global trade
system. Consumer industries are those which supply goods directly to the population as a whole. The major growth in manufacturing industries came in the 1920s,
when goods could be manufactured at a low enough price so as to create a much
larger market. The overall change has been from a very localised, 'elite' form of
consumerism to global, commercial consumerism. The global locations of
manufacturing industries today reflect this change. Ironically, the real meaning of
the word 'manu-facturing' is 'hand-made', which is certainly not the case in most
modern, so-called manufacturing industries.

e This is a very sophisticated response, though there are some odd errors, such as 'relatively insignificant in today's trading system'. Despite an obvious command of terminology, the candidate omits important aspects of what manufacturing industries actually do; specifically, there is no mention of turning raw materials into usable goods. The answer is awarded 3 marks.

(b) The original location of manufacturing industries, and certainly the car industry, was near to the raw materials essential for production, e.g. iron and coal (for burning), as transportation of the raw materials was more expensive than the finished product. In general, this trend has changed today, and industries have market locations.

There has been an overall global change in location from the more economically developed countries (MEDCs) to newly industrialised countries (NICs) and less economically developed countries (LEDCs). The shift to NICs happened about 20–30 years ago, while the shift to LEDCs is happening now. This is due to the need to exploit new markets and find cheaper labour.

It has, however, become increasingly difficult to generalise about the changing locations of manufacturing industries in today's global economy, as some places have deindustrialised and lost employment in secondary industries, e.g. the United Kingdom, while other places have fairly recently industrialised, e.g. Spain.

e The candidate presents a very strong opening set of statements about causes, but nothing on consequences. A good command of terminology and an ability to see the weaknesses of generalisation are evident.

Classical location theory was developed around Weber's 'least-cost' model, which was published in 1926. The principles of Weber's theory are based around a series of assumptions, among them that transport costs are proportional to the distance covered by a raw material or a finished product and proportionate to the weight of goods. The markets are also fixed, and perfect competition exists. The principles of this model may have been very relevant to manufacturing industries, when it was published; but manufacturing industry locations have changed, and in doing so they highlight major weaknesses in classical location theory.

When the theory was circulated, freight rates were determined by distance multiplied by weight. But most freight rates today are not concerned with distance; they are concerned with the relationship between bulk and value of goods being transported. Freight rates today are either stepped (certain charges for certain distance 'bands') or there is a uniform delivery price. This is all due to the developments of transport links and methods today. Transport costs are now so tiny (about 2–3% of overall costs) that they are practically insignificant. Unlike when the model was developed, manufacturing industries are not organised within a national framework. Industry location used to be determined almost entirely by proximity to raw materials due to these high transport costs. Transportation of goods is not such a limitation in the globalised world, which sees market locations becoming increasingly more common and more important than raw material locations.

Perhaps one of the biggest flaws of the classical location theory is the assumption of perfect competition. When the model was developed, manufacturing industries were not in so much competition, as there were many industries which had not been developed and exploited. In the car industry, the first truly global corporations were General Motors and Ford, and competition from other companies was insignificant. Today, ten major corporations dominate the car industry. Between them, they produce 75% of the world's cars. Despite the fact that competition is mainly between ten companies, it remains fierce. The theory assumes that points of demand are given and constant, but this is completely unrealistic today, as competition and demand are changing all the time, so profits constantly fluctuate, and therefore the cost of location does too. Competition and demand are heavily affected by other factors, such as advertising and the vast gulf between the rich and the poor. This means that the constant change in manufacturing industry location is not as easily explained as the classical location theory suggests.

It is almost impossible to apply the classical location theory to today's industry, due to the growing issue of globalisation. Weber suggests a very specific type of enterprise: owner-managed, single-plant and single-product. This is almost unheard of now, due to the rise of transnational corporations (TNCs) and successful national chains. For example, in 1982, GKN (an engineering firm) had three manufacturing plants and, during the recession of the late 80s, the inner-city plant had to close because that was the plant with the oldest machinery. According to Weber's theory, the plant should have been closed because of cost; but this was clearly not the case. In general, plants of a chain which have to close are not necessarily the least efficient. It is now all about internal strategy, and the pressure to modernise and globalise. The organisational structure of manufacturing industries has changed so much that the classical location theories are, to an extent, invalid and should not be considered when looking at reasons for changing location of industry.

e This is a very competent review of Weber, but the essay title is only recalled in the last sentence.

It could be considered that Fordism provided the basis for globalisation, and thus for the major changes in location of manufacturing industries, although it was first introduced by Henry Ford back in the 1920s. Fordism is a system of production, a system of demand and a new form of social institution. When first introduced, it completely reorganised certain areas of society. It emerged in a period of crisis of over-capacity, when there was a small market. The saturation of this market led to an economic depression as many workers lost their jobs and industry contracted. Henry Ford recognised the need to create a larger market, and to revolutionise car production in order to make it more efficient and profitable.

The first major breakthrough was the standardisation of parts. Ford demonstrated its use in 1912. Before this, every car had been different. Specific tasks were then set for each worker, so they became specialised. This quickly led on to the possibility of mechanisation. It was called 'Taylorisation', and removed all mental agility from the manufacturing process. At first, the workforce heavily resisted it, as workers did with the most revolutionary stage of Fordism — the introduction of assembly lines. In order to keep a good workforce in these conditions, wages were increased. Ford introduced savings plans for his workers and paid them enough to enable them to buy the cars they were making. This, and the increase in productivity which led to lower car prices, caused an increased demand, and the car industry became very profitable. The introduction of assembly lines, and mechanisation, led to productivity increases, generating more profit for the company. International location began when it was realised that Fordism could be successful in any country which had no other industry of the same kind located there.

On a national level, the original location of car industries can be demonstrated using the UK as an example. As with many early manufacturing industries, the location of the industry was determined by the location of raw materials, mainly iron, and an agglomeration effect occurred. Most firms concentrated in the West Midlands, around Coventry, Birmingham and Wolverhampton. Birmingham was the metalworking centre of the UK, due to the resources there, and there were industries already there which may have been of use to the car industry, e.g. Triumph (motorbike manufacturer), Singer (sewing machine manufacturer) and Wolsey (agricultural machinery manufacturer). The advantages of this location were reinforced when First World War engineering firms grew in the region. These reasons can all be explained from a Weberian point of view. Yet there were also signs of direct inward investment from foreign companies from a very early stage. One of the first firms to be established was Daimler, a German company, in Coventry in 1896. At this point, foreign companies were locating in the same regions as British companies, and for the same reasons.

> *e* Very impressive levels of detail are shown here with some appropriate links to location, though narrowly drawn from one industry.

Protectionism was imposed from 1918–1950 on all foreign vehicles in the UK, in order to strengthen the big companies such as Austin (based at Longbridge) and Morris (based at Cowley, Oxford). General Motors and Ford were established in 1928 and 1929, and their competition drove out the smaller companies. Location became increasingly concentrated in the West Midlands.

Throughout the century, market locations have become increasingly popular, and there has been a distinct shift southwards towards London and the M4; 85–90% of car production is now based along the M4 corridor or in London. This is due to the amount of space there for building and expansion, the amount of flat land, and the communication benefits of locating near the country's trade centre. The main changes in the UK have been from north to south, and from traditionally industrial areas to areas now known for high-tech industries. The reasons for this shift of manufacturing industries southwards include the pull of market locations, improved communication and transport links, and cheaper cost of transporting raw materials. In many cases, it is cheaper to import raw materials than use local ones, due to depletion of local resources, e.g. coal mining in south Wales is dying out while cheap imports from Australia are now popular. In recent years, the UK has also seen much inward investment from foreign corporations, e.g. Nissan, Toyota and Honda. This demonstrates the impact globalisation has had on the UK, and the entry that Japan has made into the European and US markets.

Japan was an NIC about 30 years ago. Post-war Japan was significantly different to the world-trading giant we know today. Its strategy for entry into the global markets involved importing Western goods (mainly from the US), taking them apart and improving them. Protectionism was introduced initially to ensure domestic sales, and hence the means to compete with the economies of scale of foreign TNCs. Japanese companies raised enough capital to globalise production. This is how Nissan, Honda and Toyota became global corporations. On a local scale, the industrialisation of Japan led to a significant shift in the location of most industry, and in the financial centre of the country. The shift was from the North Island to the South Island. Here, huge corporations were developed, e.g. Yokohama and Kawasaki motorbikes which come from the cities of the same name, near Tokyo. The strategy used by Japan is known as import-substitution industrialisation and this is used by many NICs, though normally with much less success than Japan since the domestic market must initially have the funds to purchase these new products.

e Again, the level of detail is excellent but the focus gets lost from time to time. It would be useful to look back at the title which is occasionally forgotten amidst the dates and places.

As a trading nation, Japan today is so successful that many high-tech industries from Europe and the US have relocated there in order to benefit from their techno-logical expertise and to be close to a market demanding these high-tech goods. A system known as 'just-in-time' (JIT) production is used by Japanese car companies, and now by many other foreign companies, as it is much more economically efficient than old, Fordist 'just-in-case' methods. The principle behind the JIT system is flexibility and involves having flexible machinery, which can be quickly adapted to new models, with workers organised to rotate jobs and respond quickly to changes. The flexible machinery and labour allow short runs of individual products. A close relationship with the whole supplier network reduces stocks to a minimum and maintains a constant flow of production. Parts are organised to arrive from suppliers 'just in time', which maintains a fast flow of production. In contrast to this system, the US 'just-in-case' system is being abandoned in favour of JIT strategies. JIT methods of production are a response to the growing global market locations of manufacturing industries, and the requirement for mass customisation rather than mass production. The change has come about due to improved technologies and the ability to customise products, and the increasingly global location of markets gives corporations a first-hand idea of market needs.

There are many reasons for the increasing globalisation of manufacturing industries and it has many effects. TNCs have an important position in all of this. The world's largest 1000 TNCs control four-fifths of the world's total output of goods and services. TNCs are becoming increasingly rich and powerful due to their low long-run average costs, which are a result of their ability to take advantage of economies of scale. With such a high capital base, they are able to finance high-risk investment and, most importantly, have access to a global market. General Motors is the world's largest corporation, with an annual turnover greater than the GDP of all but 14 countries in the world. A third of the company's output is from plants outside of the US. These facts demonstrate both the global power and wealth of the TNCs, and their global locations. TNCs are so powerful that they have almost no barriers preventing international location, thus creating a globalised market.

The globalisation of manufacturing industries is also due to the attraction of market locations. Raw materials can be very cheaply imported today; distribution costs in general are significantly lower than they were when Weber first put forward his model. This means that corporations can locate in particular countries near to their targeted markets, avoiding tariffs of protectionist economies, adapting models to suit market needs and often utilising a low-cost labour force. This is why many TNCs have chosen to locate in developing countries: they have labour forces which will not demand high wages, and there is often plenty of space to locate and expand, with the ability to charge lower prices than local companies due to their economies of scale.

It is very difficult to give firm reasons as to why certain manufacturing industries are located where they are, because there are so many contributing factors, which themselves are dynamic and complex. This is another weakness of classical location theory; it is presumed that the markets and factors affecting location are static. Reality clearly shows dynamic changes which make it difficult to be sure of location reasons without stereotyping. For example, some manufacturing industries today may still be located where they were originally, yet the reasons for that location are no longer applicable. This could be due to new innovations, e.g. cotton being replaced by synthetic fibres in the textile industry, or sources of energy, e.g. water power being replaced by electricity. Some sites were simply chosen because of the individual preference of the entrepreneur providing the capital, e.g. Rowntree's at York. Many industries also found their present locations because of trial and error, and not through any strategy or specific reason. So there is a certain amount of ambiguity in attributing reasons to locations, and this must be taken into account.

e These concluding remarks about causes add some new detail.

Without doubt, the rise of TNCs is a main cause of globalisation and changes in the location of manufacturing industries. Globalisation has broken down almost all trade barriers and provided a free, open market for the world, with the exception of a few protectionist economies. It is now increasing to the extent that examining industrial change on a national level is becoming much less significant. Location is now based on a global framework controlled by TNCs.

The global dynamic suggests that TNCs will continue to rise and that globalisation will become an increasingly important issue. No more tariffs will exist because TNCs will have the power to demand location in any country they desire. The employment and direct inward investment will attract the host country (although cheap labour is now often brought to the producer from other parts of the world). Many people in the Western world view this globalisation, and its consequences, as entirely positive because it means a constant lowering of global product prices. But this will happen at the expense of the LEDCs, which have never had the chance to develop a domestic market and so have little money to benefit from this inward investment. The people of these countries will be just as far removed from the world of trade because they will simply not be able to afford a place in it. Fordist strategies will disappear because TNCs no longer need to benefit from them, so the opportunity for the workers to purchase their own goods will disappear. The increase in TNCs will also see the disappearance of all privately owned, single-plant firms, as TNCs can exploit resources and labour with their huge wealth. This will have repercussions for the poor, since it is likely that their agricultural land and natural resources will be exploited due to the increasing use of LEDCs as locations for manufacturing industries.

e The focus is now on consequences, but it is not explicit. There is a good deal of very high level material here, though much of it is highly contentious.

So, in the future, the consequences of the change of manufacturing industry locations could be the complete domination of markets by TNCs at the expense of the world's poor, who will become poorer. The wealth gap will become irreversibly larger. However, it must be noted that in certain situations, the consequences of current changes are yet to be perceived... and will not be for years to come.

e **This is a very unusual essay. The most obvious feature is its great length. A good deal of it could have been cut without impacting on the final mark, especially the lengthy material on production methods which need not be explained in such meticulous detail when addressing location changes. It was written in an hour, leaving only 30 minutes for the planning and writing of the second essay in the examination. Nonetheless, it demonstrates an extremely sophisticated understanding of the topic and has an immensely detailed range of information about the changes in world industry over the past century. However, the range of examples is quite narrow with the car industry dominating, and the focus on the title gets lost in the detail from time to time. It is clearly a Level 5 response. Bearing in mind that examiners should award across the whole mark range and given the very high quality of this essay, the answer to part (b) is awarded 20 marks.**

Total mark: 3 + 20 = 23/25

Rural–urban interrelationships (I)

Study the photograph below, which shows a grape harvest in South Africa.

(a) **Distinguish between intensive and extensive agriculture.** (5 marks)

(b) **Examine the reasons for variation in the average size of agricultural units.** (20 marks)

■ ■ ■

Candidate's answer to question 3

(a) Intensive agriculture takes place mostly in the developed world, where there is more money to spend on machinery and equipment. It is particularly common in countries such as England and Holland, where there is a shortage of land. In these circumstances, farmers are forced to get their profit from adding more and more fertiliser in order to increase their output. In many LEDCs, there is abundant land and farmers do not have to do this, which is fortunate since they seldom have the resources to invest in this way. Intensive farming also tends to be arable.

e This is odd. The suggestion that shortage of land encourages intensification is legitimate and illustrated appropriately, but the supposition that LEDCs have abundant land is clearly wrong. The concept of yield is not addressed directly and the use of the term 'profit' reinforces the apparent conflation of extensive with subsistent, and intensive with commercial. The response is awarded 2 marks.

(b) There are many reasons why farms vary in size. This can depend on whether they are located in MEDCs or LEDCs. It also depends on the type of farm in question, e.g. intensive or extensive, and on physical and human reasons. These factors can have a positive or negative effect on the scale or size of the farm; its size can vary over time and space as well.

e This rather weak opening statement says nothing except that everything varies.

One of the key reasons why farm size varies depends on whether it is intensive or extensive. Intensive farms overall are small-scale because they wish to get the highest yield from the smallest area, with high inputs of capital and/or labour. An MEDC example would be market gardening and an LEDC example would be rice farming in the Philippines. Therefore intensive farms are on a smaller scale than extensive farms.

e It is a shame that this material was not given as the answer to part (a).

Extensive farms are not on a micro- or meso-scale like intensive farms, but on a large macro-scale. This is where there is a low input of capital and/or labour per unit area, producing lower yields. These types of farms do not depend on getting the highest yield possible per unit area. This is often to do with the type of farming, e.g. ranching in America where a large amount of land is needed. They are also often located further from the market than intensive farms, as the land is cheaper there; and, unlike dairy farming for example, they do not have to take products into the market on a daily basis.

However, the size of an intensive farm in an MEDC such as the UK can still be a lot larger than an intensive farm in an LEDC such as China. Extensive sizes also vary slightly depending on the country. This is often due to the amount of capital and government assistance available, which is found more in MEDCs than LEDCs, e.g. the EU Common Agricultural Policy and milk quotas.

e There is some detailed knowledge here, though it could be better linked to the title. The curious use of the phrase 'extensive sizes' suggests some misunderstanding again.

Farms also vary in size depending on whether they are subsistence or commercial. Subsistence farms are smaller because of where they are located and depend on whether the farmer wishes to maximise profits or not. For example, there are few subsistence farms in western Europe; subsistence farms are smaller because they are predominantly producing crops themselves and then selling the small surplus. Commercial farmers, e.g. dairy farmers in the UK, pig farmers in Denmark and cereal farmers in the Prairies, are all trying to get a maximum profit and want to sell everything they produce, not feed it to their families. To make a profit, they have to operate on a larger scale and over a larger area than subsistence farmers, who are predominantly in backward LEDCs.

e The first part of this is confusing and the logic is flawed. There is a grain of truth, but the argument is not tenable, given the existence of commercial rice farms in Japan of less than 5 hectares.

3

Commercial farms can vary in size, depending on whether they are intensive or extensive, arable or pastoral. Arable farming (e.g. dairying) is commercial, e.g. Devon and Cornwall. However, generally these farms are a lot smaller than commercial cereal farms in the USA. Therefore commercial farm sizes are dependent on where they are and what they are producing.

✎ 'Arable farming (e.g. dairying)' — is this a slip of the pen?

Farm size also varies according to whether the farm is nomadic or sedentary. A nomadic subsistence farmer in Brazil is going to be constantly moving around and the areas he uses will be small. However, a sedentary farmer in East Anglia will have a larger farm because he owns the land and does not move his farm seasonally.

✎ Here is another very confused idea.

So, the size of a farm can have a positive or negative effect. The nomadic farmer farms on a small scale and thus he is able to move, whereas a commercial farmer is restricted to the land he owns. But the commercial farmers have the advantage of an established farm.

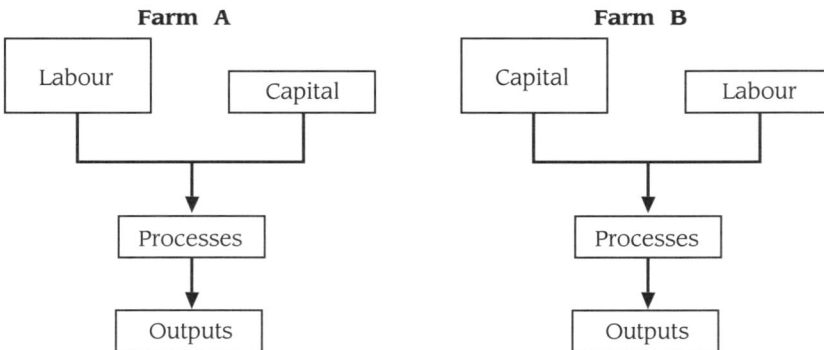

Farm A

Labour	
	Capital

Processes

Outputs

Farm B

Capital	
	Labour

Processes

Outputs

The diagram above shows why the sizes of a farm can vary. Farm A is an intensive rice farm in China, whereas Farm B is an intensive arable farm in the UK producing cereals. Both are intensive but Farm B has larger yields, due to the availability of more capital and processes.

✎ This is a useful diagram, but it has been inserted in an odd place.

Mechanisation can also help to explain why farms vary in size. MEDCs are still, overall, more advanced than LEDCs, with machinery such as tractors. This makes it possible for them to farm on a larger scale using a smaller workforce, e.g. to plough. Mechanisation helps to speed up the processes of sowing and harvesting, producing more yield and therefore capital to increase farm size.

MEDCs are also ahead in the amounts of fertilisers, pesticides and herbicides available to them to make it viable to grow on a larger scale. However, the Green

Revolution is making it possible for LEDCs to increase farming yields through hybrids. They are now also being given the advantages of fertilisers and other such benefits enjoyed by MEDCs, again causing them to improve yields. This leads to a positive spiral of growth in farm size.

Physical factors, such as climate, are another important reason why farm sizes vary. They can be both positive and negative. For example, a wet climate with large amounts of precipitation will cause farms to locate there on a larger scale than those areas with little precipitation. Over time, however, irrigation schemes will make it possible to locate in other areas.

e This is a brave attempt, but it does not say enough to be meaningful. Again, there is some confusion over terminology: 'locate there on a larger scale' is a very ambivalent phrase.

Soil type is important. A farmer is more likely to locate on a fertile soil than on one which needs lots of fertilisers. If the soil is permeable or impermeable, this may be a factor. The relief of the land also explains why farms vary in size. Farmers prefer flat land as it is easier to farm than steep gradients, e.g. the flat land of Kent and Norfolk. However, steep relief is more labour intensive and needs terracing etc. which explains why farms on steep reliefs are often smaller than on flat relief.

Governments are key factors. They can promote farms, e.g. through common agricultural policies, milk quotas, and the Green Revolution in LEDCs. They can also have a negative effect by reducing farm size, e.g. set-aside farms. They can provide grants and loans to help farms expand. Negatively, they can implement strict controls, e.g. on the amount of hedgerows and environmentally special areas (ESAs) that can be built on.

e Set-aside does not reduce the size of the farm.

Therefore farm size is affected by many factors over time, positively and negatively. It depends on farm type, where it is, and physical and human factors. It is not as simple as von Thunen believed.

e **There is no question that this response to part (b) covers a great deal of ground. However, the approach is sweeping with only occasional local detail and many generalisations. These are sometimes wrong and often confused. It appears that the key advice about writing essays has been remembered, as in the need to contrast at different scales, include LEDC and MEDC comparisons whenever possible and remember that most changes have both negative and positive aspects. This is a Level 3 essay, worth 10 marks.**

Total mark: 2 + 10 = 12/25

Question 4

Rural–urban interrelationships (II)

Study the graph below, which shows the estimated and projected urban and rural population of the more and less developed countries, 1950–2030.

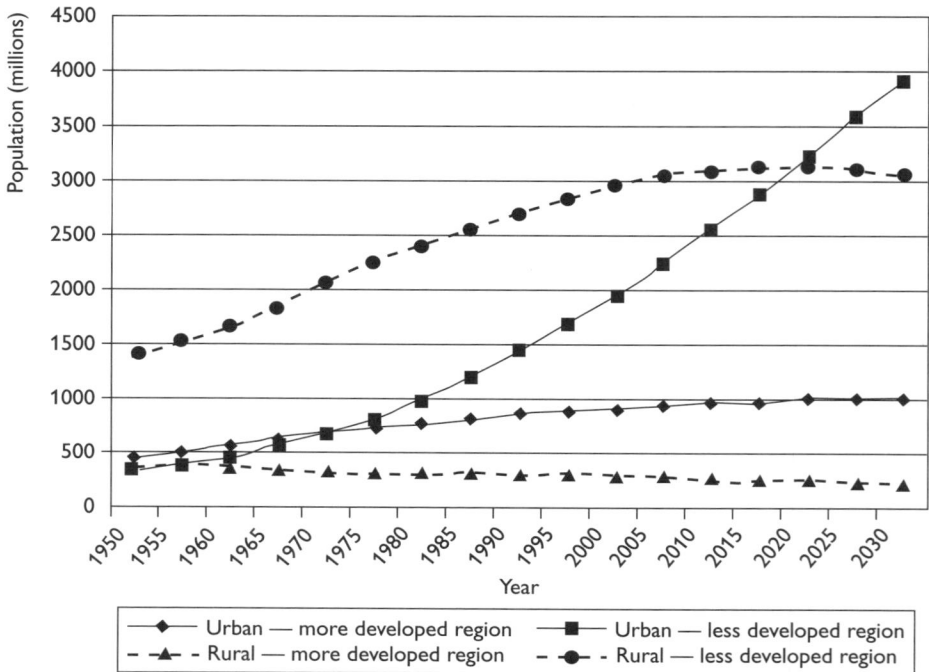

(a) **Define the term 'urbanisation'.** (5 marks)

(b) **Examine the causes of variations in the rate of urban growth in the modern world.** (20 marks)

■ ■ ■

Candidate's answer to question 4

(a) Urbanisation is a process that takes place when cities grow. It can be measured in terms of either numbers of people or the area of the city. London has grown a great deal throughout history and is now a city of more than 7 million people. In the last 20 years it has stopped growing and has got a little smaller in population, though not in area. People have moved out of the city to live in the surrounding countryside which they see as being more peaceful, with lower crime rates. In LEDCs people are still arriving in cities; more than a thousand a day arrive in São Paulo, for example.

e This is a very sound definition that covers population and area. Both measures are half-truths, but there is an attempt to exemplify. There is no sense here of urbanisation as an economic process, and the cliché about country living is not helpful. It is awarded 3 out of 5 marks.

(b) There are many reasons why cities have grown and these reasons have changed over time and from place to place. The most important reason for the existence of cities is the many functions that they perform. In some places these have changed, which helps to explain why cities have grown at different rates in different places. London grew most rapidly during the Industrial Revolution when it was a major port and home to many industries like furniture making and sugar refining. The East End was the centre of this industrial zone, and a large number of people came to the city in order to work in these industries. At the same time, people were leaving the countryside because of the many push factors like poverty and poor facilities. London grew very rapidly in the Victorian period. It was a centre of government and as people started to earn more money, it also developed an important retail centre around Oxford Street and Regent Street. It was the capital city and this attracted civil servants and politicians. As a result of these functions, London grew very rapidly, becoming the largest city in the world by the end of the nineteenth century. Being the centre of the Empire obviously helped this growth.

ℓ The first four lines provide an excellent example of how to fill space without actually saying anything. It would be far better to omit this and just get on with the essay, or offer a view of what might be the dominant reasons or the dominant functions.

Since then, there have been changes in the rate of urbanisation. In the case of London, it has fallen recently. This has been because many people have moved out of the city. The main reason has been counter-urbanisation, with the richer members of society deciding to move to rural areas and commute into the city instead. They have left the inner-city areas in particular, such as Kentish Town and Camden, moving out to counties such as Oxfordshire. This has had a big impact on both the areas that they have left and the areas to which they have moved. Many of them have found it difficult to be accepted in the rural areas.

Another reason for the falling population is that some of the industries which used to be in the East End of London have disappeared. Some have moved out to rural areas: the furniture industry, mentioned earlier, has moved out to High Wycombe; sugar refining, which used to be at Tate and Lyle's factory on the River Thames, is now situated in the sugar beet fields of places like Norfolk. Because these jobs have gone, so have the people, who now live in other parts of the country. New jobs have developed in different places and some of these have had quite rapid growth. Reading is an example of this, with jobs in hi-tech industries like computers offering employment that has, of course, attracted many people to the area. There are some places in England that have grown quite rapidly, mostly in the South. Other cities in the North, like Sheffield, have lost their heavy industries completely. In the case of Sheffield, the iron and steel industry has gone to countries like Taiwan and Korea, where the costs of labour are much cheaper. This has led to the growth of cities in these places.

In other LEDCs, there has been very little industry and the cities have grown because of push factors rather than pull factors. These push factors are mostly to do with changes in farming, which have led to lots of peasants having to leave the land if they cannot get jobs in the new commercial farms that have grown in places like Egypt. Cairo has grown very fast because of this rural–urban migration, and the government has tried to stop it happening by developing rural areas with large projects. This type of urbanisation is a major problem in many developing countries where squatter settlements are very common. These 'favelas' are usually without any form of sanitation and often have no proper sewage disposal. As a result, many people suffer from easily avoidable diseases like malaria, and there is much malnutrition. Jobs are rare, with many having to work informally by selling things in the street or working at home for very low wages. Most governments do not have enough money to help much by putting in better services, and there is often a lot of corruption. Some Western governments do not think that there is much point in trying to help these very poor countries which are often in Africa, and believe it is better to solve the urban problems of richer LEDCs.

e This is a seamless account with lots of good, accurate detail, though expressed rather clumsily. There is some simplification as in 'Sheffield to Taiwan', but the thrust is appropriate.

Urbanisation is still a very important trend in the world, though it has slowed down. It varies from place to place because some countries like Britain have urbanised almost completely, while others like Egypt are just starting to develop and will go on urbanising for quite a long time into the next century. In the past, the most important single reason for urbanisation was industrialisation, but that is not true today. In Africa, many cities like Cairo are growing very fast, though there is little industry. Here, the reasons are more to do with what is going on in rural areas, where there is increasing poverty and loss of land. Meanwhile, in the MEDCs, cities are stagnating as industries move out to new settlements. Only a few like Los Angeles are growing.

e There is an attempt at synthesis here, but some new material is introduced right at the end that is left unexplained. Some of this is repetitive rather than conclusive.

e **The essay as a whole is not well constructed, with many changes of direction. It might have benefited from a plan, especially to organise the middle section. However, the material is relevant, there is good use of examples (the material on London is a little dated but appropriate nonetheless) and a proper attempt is made to explain some of the complexities and to qualify the generalisations. The answer to part (b) is awarded 15 marks out of 20.**

Total mark: 3 + 15 = 18/25

Development processes (I)

Study the cartoon below about global relationships.

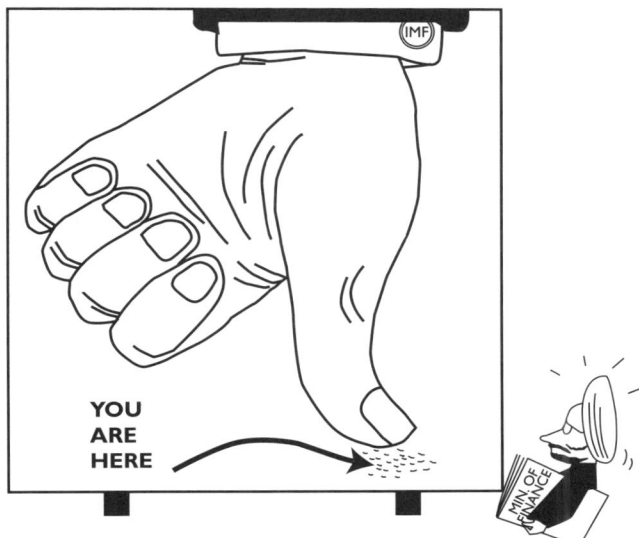

(a) **What is the global economy?** (5 marks)

(b) **How and why are countries economically interdependent?** (20 marks)

■ ■ ■

Candidate's answer to question 5

(a) The global economy is an interrelated network of economies and trade in which there is unrestricted movement of capital, workforce and products. It supersedes the national economies of members signed up to the World Trade Organisation (WTO) due to the inability of member states to impose import tariffs or block imports. The global economy has enabled transnational corporations (TNCs) to establish raw material extraction in peripheral countries, where extraction and labour are cheap, and manufacturing in semi-peripheral countries, where the relevant skills are found but where they are not as expensive as in developed countries, where the headquarters and quaternary sectors are established, along with the market. For example, the cost of labour in Mexico is one-sixth of US labour; however, when selling to an American market, the prestige of an American headquarters is very important to consumer goods purchasers.

🖉 This starts extremely well, but gets distracted by the impact of these changes. Terminology is handled well (e.g. peripheral, semi-peripheral) and the division of labour is identified as a major component. It is a shame that capital flows are ignored after the initial statement. It is awarded 4 out of 5 marks.

(b) In the modern world most countries are economically interdependent. This means that they rely on imports from other countries and on the ability to export to other countries to survive in the 21st century. It is now almost impossible for a country to survive on an autarchic basis, with only large countries such as the United States of America having the possibility of achieving this. However, during the last century, it was possible for Russia and China to maintain near-independence from trade, as did some other socialist states. With the increasing rise of the global economy and transnational corporations, the possibility of this is being removed. Countries are economically independent for a number of reasons and in a variety of ways.

> *e* There are lots of strongly assertive statements here, with a confident handling of autarchy.

The World Trade Organisation (WTO) is one of the key factors linking countries economically. Its member states are not allowed to impose tariffs on or block the import of goods from other countries in the attempt to establish a world of free trade and thus a global economy. The interrelationship of nations enables developed countries to gain cheap raw materials, such as coffee from Colombia, to sell in their country while exporting specialist manufactured goods, such as washing machines, to the peripheral countries. The peripheral countries are gaining foreign currency through export of goods to the developed world. Many non-governmental organisations (NGOs) are now arguing that free trade is unfair, because it enables rich countries, such as the USA, to become richer at the expense of the poor countries, such as Peru. This is because it is impossible for the less economically developed country (LEDC) to establish its own specialist industry, such as motor cars, because without trade tariffs the cost of their national car would be far higher than the imports of Japanese cars, for example. Under the WTO, the LEDCs are powerless to prevent this.

> *e* The tariff argument is handled well and the idea of comparative advantage is introduced, though not explicitly. Exemplification is good.

Many former colonial countries retain strong trade links with their former administrators. This is because there was an established trade link that benefited the administrators through cheap imports of goods, such as spices from India to Britain. The new trade treaties benefit the former colonies because they enable them to gain foreign currency to import goods from other, more economically developed countries (MEDCs), such as cheap agricultural goods from the USA, to establish their national economies. The rise of TNCs has greatly strengthened the interrelations of countries in the world economy. TNCs have strengthened the links between countries because it benefits them to be able to move capital freely as well as reduce trade tariffs. It is necessary to remember that in terms of turnover, some TNCs are more powerful than national governments, and if Microsoft were a country, it would have the fifteenth highest gross national product (GNP). TNCs have strengthened countries' links with one another because it is cheap for them to employ sweatshop-style labour in peripheral countries like Korea, and then export goods to the Western developed market. For example, Microsoft manufactures 75% of its computers in Ireland, a semi-peripheral country. This is because transport costs — both haulage and terminal costs — are only about 1–3% of annual expenditure, while in EU countries labour costs can be 10–40% of expenditure. TNCs maintain the quaternary and tertiary sectors in MEDCs close to the market for prestige reasons; however, in manufacturing, Weber's model of individual location is dominant.

e The general approach is excellent, though some of the detail is contentious (is Ireland really semi-peripheral?) or wrong (is Korea peripheral?). The level of economic understanding is very impressive.

Countries are economically interdependent for a variety of reasons, the most important of which is their membership of organisations such as the WTO. However, this membership may be reduced in the future if LEDCs are influenced by the NGOs calling for an end to globalisation, as the Seattle riots in 1999 showed. As the influence of colonialism decreases, it is likely that trade will become even more global, with nations trading with more than one partner country. However, as TNCs continue to grow, as the $200 billion merger to form America On Line (AOL) showed, world trade will increase.

e The conclusion is a trifle disappointing in the light of what has come before. Membership of the WTO is not compulsory and the candidate had already dealt effectively with the economic argument for global trade. The final comment is not conclusive and suggests that the student ran out of time.

e **The response to part (b) as a whole is an effective piece of writing, showing a command of the key ideas and an ability to express them and illustrate them clearly. The answer is well focused on the question and the argument is sustained effectively. It is awarded 18 marks.**

Total mark: 4 + 18 = 22/25

Development processes (II)

Study the table below, which shows average house prices in regions of the UK.

Regional statistics — average dwelling price			
(£) 1999 UK	**92 521**		
Northeast	61 620	England	96 133
Northwest	73 509	Wales	67 483
Yorkshire and the Humber	67 416	Scotland	69 312
East Midlands	72 437	Northern Ireland	66 267
West Midlands	79 757		
East	68 410		
London	142 321		
Southeast	121 654		
Southwest	89 217		

(a) What do you understand by cumulative causation? (5 marks)

(b) Why are some regions richer than others? (20 marks)

■ ■ ■

Candidate's answer to question 6

(a) Cumulative causation is closely related to the multiplier effect in that, in brief, it covers the idea that once economic growth begins, it tends to gather pace through the circulation of money and, economically, one thing leads to another. Thus an initial investment of capital in industry (whether by government or by private individual) is multiplied because workers spend money in the local community and the traders who receive this money spend it themselves, and so on. Similarly, an industry demands goods from component suppliers and they in turn stimulate demand from other suppliers. People pay more taxes and the local services improve, making growth regions even more attractive. This formed a key part of Gunnar Myrdal's work on regional economics when he tried to explain why regional disparities tend to persist. He won the Nobel prize for his contribution.

🖉 This is a very good answer, with some biographical detail. It is awarded 5 marks.

(b) According to classical equilibrium theory, all areas should have equal wealth. As firms go to areas of lower wages (particularly those firms that have low barriers to movement, e.g. multinationals and TNCs), they employ people until labour gets more expensive and workers demand higher wages. They then leave to go to another area of lower wages. As costs increase, speed of movement depends on sunk capital. Therefore, the wealth of regions becomes equal as all have the same wage rates.

🖉 The expression here is clumsy, though the point does emerge. The role of TNCs is not really an issue in this context.

An idea developed by Rostow is that regions develop around natural resources, as this is where industry sets up, creating wealth and jobs. However, an area of great wealth in the 1920s was south Wales, where the resources of coal and iron ore meant that small towns like Dowlais and Ebbw Vale were some of the richest areas. However, this is no longer the case as the area has become a supply region with a small pool of unskilled manual labour. Similarly, the Netherlands, which is a very rich region, developed the factory system but has no resources. Japan also has few resources but has achieved the greatest growth over the last 40 years, regularly achieving growth of 10–12% of GDP. The two wealthiest regions in resource terms are the US and Brazil, but these are not the two richest regions, with the US at the top end of the scale and Brazil at the bottom end.

e This is not Rostow's model (though remember that you cannot lose marks so it does not affect the quality of the comment in a general sense). The material on south Wales would be stronger if some data were offered.

It is for the reason above that physical resources have to be dismissed as a reason for differences in wealth between regions.

e This is a very bold assertion. It would be improved if it was qualified by suggesting either that natural resources can be very helpful in economic development or that, in some cases, over-reliance on raw material extraction can lead to dependency.

An alternative theory is Myrdal's idea of cumulative causation.

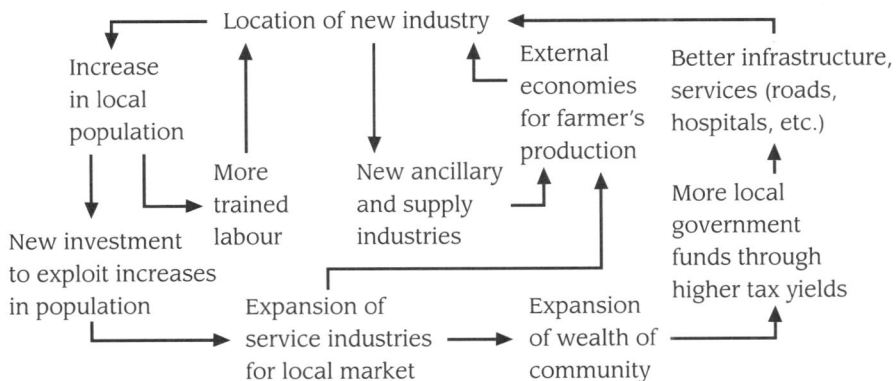

e This is an impressive feat of recall and certainly relevant.

This shows how a region can grow on the back of the location of a new industry and how the small initial disparity can be rapidly magnified. It also shows backwash effects where the costs of production in the 'core' area are lower than the 'periphery' due to economics of scale in the core. This means that the periphery cannot develop and becomes a supply region feeding the core, furthering the disparity in wealth between core and periphery. This occurs on a number of scales, e.g. the North being the periphery in England and the South being the core. On a global scale, the MEDCs are the core, extracting resources from LEDCs by multinationals and TNCs.

e The use of the diagram is good and the reference to backwash effects shows a high level of understanding.

Trickle-down (where wealth travels downwards) appears to be very slow, e.g. in Latin America as a supply region and in the Gezira cotton-growing region of the Sudan, both of which have had negative growth in recent years.

e Neither 'trickle-down' nor 'supply region' are explained and it would have been helpful to do so here.

The method by which the initial disparity develops and becomes further magnified is part of the explanation for difference in wealth or richness between regions.

Government intervention is one way in which a disparity develops — for example, by starting growth poles from which cumulative causation can start. Examples of this on different scales are the Hoover Dam and Ebbw Vale, where garden projects in 1985 were used as a method to regenerate the regions and attract firms (particularly multinationals and TNCs) to what was a declining area. Similarly, regional policies which give tax breaks to new firms may create an initial difference.

e This is a rather odd attempt to cover growth pole ideas. Nothing is said about the success of these schemes.

Accessibility is important to the new firm locations. This is a possible reason why Devon and Cornwall, being on a peninsula, have not grown as rapidly as the rest of southern Britain. They are not on any route, so regional disparity in terms of wealth occurs. However, the building of a bridge over the Humber did little to change the difference in wealth on either side of the river, showing very small-scale disparity.

e There is an attempt to change scale here, which is pleasing.

Climatic factors have an impact. For example, the three regions that show greatest growth in research and development are the UK, the Mediterranean and the US (sunbelt-California). As competition for workforce is very competitive, they have set up in ideal climates. This is why Toulouse has shown great growth with cumulative causation of research and development on the back of climate. Similarly, Seattle is Microsoft's centre, and climate has allowed an initial disparity between regions to be created as people are attracted to relocate in the region because of the temperate climatic conditions, with no extremes of climate.

e This essay ends very abruptly with no attempt to draw the strands together. There is a good deal of summary comment in the text itself, however.

e **Overall, the response to part (b) is something of a curiosity. There is an impressive command of theory laced with some exemplification. The approach is combative and a view is certainly taken. The organisation is poor, which spoils the overall impact because it is hard to follow the logic. It is awarded 14 marks out of 20.**

Total mark: 5 + 14 = 19/25